Science and Philosophy

ALFRED NORTH WHITEHEAD

The twentieth century has produced few men whose achievements in philosophy can surpass those of Alfred North Whitehead. His is not a mere technical excellence. It is a competence which, on occasion is adorned by an unexcelled brilliance of vivid expression. Learning sits lightly on his firm shoulder. A sparkling, somewhat ironic, humor shines through his discourse. The profound humility of the truly wise dignifies his utterances. To persons in all walks of life his is a rare exemplification of cultured charm and dignity. The essays contained in the present volume represents a cross-section from the career of the distinguished philosopher. Some are of biographical nature, as the enchanting "Memories"; others treat of the perennial philosophical problems. Here are, also, Whitehead's thoughts on the meaning and future of learning.

SCIENCE
AND PHILOSOPHY

Science
and Philosophy

ALFRED NORTH WHITEHEAD
O.M., F.R.S., F.B.A., D.SC., HON. LL.D.

*Late Emeritus Professor at Harvard; formerly Professor
at Imperial College of Science and Technology in Kensington
and Dean of the Faculty of Science in University of London*

PHILOSOPHICAL LIBRARY

New York

Contents

CONTENTS

SCIENCE

AND PHILOSOPHY

PART I

PERSONAL

Autobiographical Notes

I WAS BORN IN 1861, February 15, at Ramsgate in the Isle of Thanet, Kent. The family, grandfather, father, uncles, brothers engaged in activities concerned with education, religion and Local Administration: my grandfather, born of yeoman stock in Isle of Sheppey, was probably a descendant of the Quaker George Whitehead, whom George Fox in his *Journal* mentions as living there in the year 1670. In the year 1815, my grandfather, Thomas Whitehead, at the age of twenty-one, became head of a private school in Ramsgate, Isle of Thanet, to which my father, Alfred Whitehead, succeeded at the correspondingly early age of twenty-five, in the year 1852. They were, both of them, most successful schoolmasters, though my grandfather was by far the more remarkable man.

About 1860 my father was ordained as a clergyman of the Anglican Church; and about 1866 or 1867 he gave up his school for clerical duty, first in Ramsgate, and later in 1871 he was appointed Vicar of St. Peters Parish, a large district mostly rural, with its church about two or three miles from Ramsgate. The North Foreland belongs to the parish. He remained there till his death in 1898.

He became influential among the clergy of East Kent, occupying the offices of Rural Dean, Honorary Canon of Canterbury, and Proctor in Convocation for the Diocese. But the central fact of his influence was based on his popularity with the general mass of the population in the Island. He never lost his interest in education, and daily visited his three parochial schools, for infants, for girls, and for boys. As a small boy, before I left home for school in 1875, I often accompanied him. He was a man with local interests and influence; apart from an understanding of such provincial figures, the social and political history of England in the nineteenth century cannot be comprehended. England was governed by the influence of personality: this does not mean "intellect."

My father was not intellectual, but he possessed personality. Archbishop Tait had his summer residence in the parish, and he and his family were close friends of my parents. He and my father illustrated the survival of the better (and recessive) side of the eighteenth century throughout its successor. Thus, at the time unconsciously, I watched the history of England by my vision of grandfather, father, Archbishop Tait, Sir Moses Montefiore, the Pugin family, and others. When the Baptist minister in the parish was dying, it was my father who read the Bible to him. Such was England in those days, guided by local men with strong mutual antagonisms and intimate community of feeling. This vision was one source of my interest in history, and in education.

Another influence in the same direction was the mass of archæological remains with their interest and beauty. Canterbury Cathedral with its splendour and its memories was sixteen miles distant. As I now write I can visualize the very spot where Becket fell A.D. 1170, and can recall my reconstruction of the incident in my young imagination. Also there is the tomb of Edward, The Black Prince (died A.D. 1376).

But closer to my home, within the Island or just beyond its borders, English history had left every type of relic. There stood the great walls of Richborough Castle built by the Romans, and the shores of Ebbes Fleet where the Saxons and Augustine landed. A mile or so inland was the village of Minster with its wonderful

Abbey Church, retaining some touches of Roman stone-work, but dominated by its glorious Norman architecture. On this spot Augustine preached his first sermon. Indeed the Island was furnished with Norman, and other medi-æval churches, built by the Minster monks, and second only to their Abbey. My father's church was one of them, with a Norman nave.

Just beyond Richborough is the town of Sandwich. At that time it retained the sixteenth and seventeenth centuries, with its Flemish houses lining the streets. Its town-records state that in order to check the silting up of the harbour, the citizens invited skilful men from the Low Countries—"cunning in waterworks." Unfortu-nately they failed, so that the town remained static from that period. In the last half century, it has been revived by a golf course, one of the best in England. I feel a sense of profanation amidst the relics of the Romans, of the Saxons, of Augustine, the mediæval monks, and the ships of the Tudors and the Stuarts. Golf seems rather a cheap ending to the story.

At the age of fourteen, in the year 1875, I was sent to school at Sherborne in Dorsetshire, at the opposite end of southern England. Here the relics of the past were even more obvious. In this year (1941) the school is to celebrate its twelve-hundredth anniversary. It dates from St. Aldhelm, and claims Alfred the Great as a pupil. The school acquired the monastery buildings, and its grounds are bounded by one of the most magnificent Abbeys in existence, with tombs of Saxon princes. In my last two years there the Abbots' room (as we be-lieved) was my private study; and we worked under the sound of the Abbey bells, brought from the Field of The Cloth of Gold by Henry VIII.

I have written thus far in order to show by example how the imaginative life of the southern English pro-fessional class during the last half of the nineteenth cen-tury was moulded. My own experience was not in the least bit exceptional. Of course details differ, but the type was fairly uniform for provincial people.

This tale has another reference to the purpose of this

slight autobiography. It shows how historical tradition is handed down by the direct experience of physical surroundings.

On the intellectual side, my education also conformed to the normal standard of the time. Latin began at the age of ten years, and Greek at twelve. Holidays excepted, my recollection is that daily, up to the age of nineteen and a half years, some pages of Latin and Greek authors were construed, and their grammar examined. Before going to school pages of rules of Latin grammar could be repeated, all in Latin, and exemplified by quotations. The classical studies were interspersed with mathematics. Of course, such studies included history—namely, Herodotus, Xenophon, Thucydides, Sallust, Livy, and Tacitus. I can still feel the dullness of Xenophon, Sallust, and Livy. Of course we all know that they are great authors; but this is a candid autobiography.

The others were enjoyable. Indeed my recollection is that the classics were well taught, with an unconscious comparison of the older civilization with modern life. I was excused in the composition of Latin Verse and the reading of some Latin poetry, in order to give more time for mathematics. We read the Bible in Greek, namely, with the Septuagint for the Old Testament. Such Scripture lessons, on each Sunday afternoon and Monday morning, were popular, because the authors did not seem to know much more Greek than we did, and so kept their grammar simple.

We were not overworked; and in my final year my time was mostly occupied with duties as Head of the School with its responsibility for discipline outside the class-rooms, on the Rugby model derived from Thomas Arnold, and as Captain of the Games, chiefly cricket and football, very enjoyable but taking time. There was however spare time for private reading. Poetry, more especially Wordsworth and Shelley, became a major interest, and also history.

My university life at Trinity College, Cambridge, commenced in the autumn of 1880; and, so far as residence is concerned, continued without interruption until the summer of 1910. But my membership of the College, first as "scholar" and then as "fellow," continues unbroken. I cannot exaggerate my obligation to the University of Cambridge, and in particular to Trinity College, for social and intellectual training.

The education of a human being is a most complex topic, which we have hardly begun to understand. The only point on which I feel certain is that there is no widespread, simple solution. We have to consider the particular problem set to each institution by its type of students, and their future opportunities. Of course, for the moment and for a particular social system, some forms of the problem are more widespread than others —for instance, the problem now set to the majority of State Universities in the U.S.A. Throughout the nineteenth century, the University of Cambridge did a brilliant job. But its habits were adapted to very special circumstances.

The formal teaching at Cambridge was competently done, by interesting men of first-rate ability. But courses assigned to each undergraduate might cover a narrow range. For example, during my whole undergraduate period at Trinity, all my lectures were on mathematics, pure and applied. I never went inside another lecture room. But the lectures were only one side of the education. The missing portions were supplied by incessant conversation, with our friends, undergraduates, or members of the staff. This started with dinner at about six or seven, and went on till about ten o'clock in the evening, stopping sometimes earlier and sometimes later. In my own case, there would then follow two or three hours' work at mathematics.

Groups of friends were not created by identity of subjects for study. We all came from the same sort of school, with the same sort of previous training. We discussed everything—politics, religion, philosophy, literature—with a bias toward literature. This experience led to a large amount of miscellaneous reading. For example, by the time that I gained my fellowship in 1885 I nearly knew by heart parts of Kant's *Critique of Pure Reason*.

Now I have forgotten it, because I was early disenchanted. I have never been able to read Hegel: I initiated my attempt by studying some remarks of his on mathematics which struck me as complete nonsense. It was foolish of me, but I am not writing to explain my good sense.

Looking backwards across more than half a century, the conversations have the appearance of a daily Platonic dialogue. Henry Head, D'Arcy Thompson, Jim Stephen, the Llewellen Davies brothers, Lowes Dickinson, Nat Wedd, Sorley, and many others—some of them subsequently famous, and others, equally able, attracting no subsequent public attention. That was the way by which Cambridge educated her sons. It was a replica of the Platonic method. The "Apostles" who met on Saturdays in each other's rooms, from 10 P.M. to any time next morning, were the concentration of this experience. The active members were eight or ten undergraduates or young B.A.'s, but older members who had "taken wings" often attended. There we discussed with Maitland, the historian, Verrall, Henry Jackson, Sidgwick, and casual judges, or scientists, or members of Parliament who had come up to Cambridge for the weekend. It was a wonderful influence. The club was started in the late 1820's by Tennyson and his friends. It is still flourishing.

My Cambridge education with its emphasis on mathematics and on free discussion among friends would have gained Plato's approval. As times changed, Cambridge University has reformed its methods. Its success in the nineteenth century was a happy accident dependent on social circumstances which have passed away—fortunately. The Platonic education was very limited in its application to life.

In the autumn of 1885, the fellowship at Trinity was acquired, and with additional luck a teaching job was added. The final position as a Senior Lecturer was resigned in the year 1910, when we removed to London.

In December, 1890 my marriage with Evelyn Willoughby Wade took place. The effect of my wife upon my outlook on the world has been so fundamental that it must be mentioned as an essential factor in my philosophic output. So far I have been describing the narrow English education for English professional life. The prevalence of this social grade, influencing the aristocrats above them, and leading the masses below them, is one of the reasons why the England of the nineteenth century exhibited its failures and successes. It is one of the recessive factors of national life which hardly ever enters into historical narrative.

My wife's background is completely different, namely military and diplomatic. Her vivid life has taught me that beauty, moral and æsthetic, is the aim of existence; and that kindness, and love, and artistic satisfaction are among its modes of attainment. Logic and Science are the disclosure of relevant patterns, and also procure the avoidance of irrelevancies.

This outlook somewhat shifts the ordinary philosophic emphasis upon the past. It directs attention to the periods of great art and literature, as best expressing the essential values of life. The summit of human attainment does not wait for the emergence of systematized doctrine, though system has its essential functions in the rise of civilization. It provides the gradual upgrowth of a stabilized social system.

Our three children were born between 1891 and 1898. They all served in the First World War: our eldest son throughout its whole extent, in France, in East Africa, and in England; our daughter in the Foreign Office in England and Paris; our youngest boy served in the Air Force: his plane was shot down in France with fatal results, in March, 1918.

For about eight years (1898-1906) we lived in the Old Mill House at Grantchester, about three miles from Cambridge. Our windows overlooked a mill pool, and at that time the mill was still working. It has all gone now. There are two mill pools there; the older one, about a couple of hundred yards higher up the river, was the one mentioned by Chaucer. Some parts of our house were very old, probably from the sixteenth century. The whole spot was intrinsically beautiful and was filled with reminiscences, from Chaucer to Byron and Wordsworth. Later on another poet, Rupert Brooke, lived in the neighbouring house, the Old Vicarage. But that was after our time and did not enter into our life. I must mention the Shuckburghs (translator of Cicero's letters) and the William Batesons (the geneticist) who also lived in the village and were dear friends of ours. We owed our happy life at Grantchester to the Shuckburghs, who found the house for us. It had a lovely garden, with flowering creepers over the house, and with a yew tree which Chaucer might have planted. In the spring nightingales kept us awake, and kingfishers haunted the river.

My first book, *A Treatise on Universal Algebra,* was published in February, 1898. It was commenced in January, 1891. The ideas in it were largely founded on Hermann Grassmann's two books, the *Ausdehnungslehre* of 1844, and the *Ausdehnungslehre* of 1862. The earlier of the two books is by far the most fundamental. Unfortunately when it was published no one understood it; he was a century ahead of his time. Also Sir William Rowan Hamilton's *Quaternions* of 1853; and a preliminary paper in 1844, and Boole's *Symbolic Logic* of 1859, were almost equally influential on my thoughts. My whole subsequent work on Mathematical Logic is derived from these sources. Grassmann was an original genius, never sufficiently recognized. Leibniz, Saccheri, and Grassmann wrote on these topics before people could understand them, or grasp their importance. Indeed poor Saccheri himself failed to grasp what he had achieved, and Leibniz did not publish his work on this subject.

My knowledge of Leibniz's investigations was entirely based on L. Couturat's book, *La Logique de Leibniz,* published in 1901.

This mention of Couturat suggests the insertion of two other experiences connected with France. Élie Halévy, the historian of England in the early nineteenth century, frequently visited Cambridge, and we greatly enjoyed our friendship with him and his wife.

The other experience is that of a Congress on Mathematical Logic held in Paris in March, 1914. Couturat was there, and Xavier Léon, and (I think) Halévy. It was crammed with Italians, Germans, and a few English including Bertrand Russell and ourselves. The Congress was lavishly entertained by various notables, including a reception by the President of the Republic. At the end of the last session, the President of the Congress congratulated us warmly on its success and concluded with the hope that we should return to our homes carrying happy memories of "La Douce France." In less than five months the First World War broke out. It was the end of an epoch, but we did not know it.

The *Treatise on Universal Algebra* led to my election to the Royal Society in 1903. Nearly thirty years later (in 1931) came the fellowship of the British Academy as the result of work on philosophy, commencing about 1918. Meanwhile between 1898 and 1903, my second volume of Universal Algebra was in preparation. It was never published.

In 1903 Bertrand Russell published *The Principles of Mathematics*. This was also a "first volume." We then discovered that our projected second volumes were practically on identical topics, so we coalesced to produce a joint work. We hoped that a short period of one year or so would complete the job. Then our horizon extended and, in the course of eight or nine years, *Principia Mathematica* was produced. It lies outside the scope of this sketch to discuss this work. Russell had entered the University at the beginning of the eighteen nineties. Like the rest of the world, we enjoyed his brilliance, first as my pupil and then as a colleague and friend. He was a great factor in our lives, during our Cambridge period. But our fundamental points of view—philosophical and sociological—diverged, and so with different interests our collaboration came to a natural end.

At the close of the University session, in the summer of 1910, we left Cambridge. During our residence in London, we lived in Chelsea, for most of the time in Carlyle Square. Wherever we went, my wife's æsthetic taste gave a wonderful charm to the houses, sometimes almost miraculously. The remark applies especially to some of our London residences, which seemed impervious to beauty. I remember the policeman who saw a beautiful girl let herself into our house in the early hours after midnight. She had been presented at Court and had then gone to a party. The policeman later enquired of our maid whether he had seen a real person or the Virgin Mary. He could hardly believe that a real person in a lovely dress would be living there. But inside there was beauty.

During my first academic session (1910-1911) in London I held no academic position. My *Introduction to Mathematics* dates from that period. During the sessions from 1911 to the summer of 1914, I held various positions at University College, London, and from 1914 to the summer of 1924 a professorship at the Imperial College of Science and Technology in Kensington. During the later years of this period I was Dean of the Faculty of Science in the University, Chairman of the Academic Council which manages the internal affairs concerned with London education, and a member of the Senate. I was also Chairman of the Council which managed The Goldsmith's College, and a member of the Council of the Borough Polytechnic. There were endless other committees involved in these positions. In fact, participation in the supervision of London education, University and Technological, joined to the teaching duties of my professorship at the Imperial College constituted a busy life. It was made possible by the marvellous efficiency of the secretarial staff of the University.

This experience of the problems of London, extending for fourteen years, transformed my views as to the problem of higher education in a modern industrial civilization. It was then the fashion—not yet extinct—to take a narrow view of the function of Universities. There were the Oxford and Cambridge type, and the German type. Any other type was viewed with ignorant contempt. The seething mass of artisans seeking intellectual enlightenment, of young people from every social grade craving for adequate knowledge, the variety of problems thus introduced—all this was a new factor in civilization. But the learned world is immersed in the past.

The University of London is a confederation of various institutions of different types for the purpose of meeting this novel problem of modern life. It had recently been remodelled under the influence of Lord Haldane, and was a marvellous success. The group of men and women—business men, lawyers, doctors, scientists, literary scholars, administrative heads of departments—who gave their time, wholly or in part, to this new problem of education were achieving a much needed transformation. They were not unique in this enterprise: in the U.S.A. under different circumstances analogous groups were solving analogous problems. It is not too much to say that this novel adaptation of education is one of the factors which may save civilization. The nearest analogy is that of the monasteries a thousand years earlier.

The point of these personal reminiscences is the way in which latent capabilities have been elicited by favourable circumstances of my life. It is impossible for me to judge of any permanent value in the output. But I am aware of the love, and kindness, and encouragement by which it was developed.

To turn now to another side of life, during my later years at Cambridge, there was considerable political and academic controversy in which I participated. The great question of the emancipation of women suddenly flared up, after simmering for half a century. I was a member of the University Syndicate which reported in favour of equality of status in the University. We were defeated, after stormy discussions and riotous behaviour on the part of students. If my memory is correct, the date was about 1898. But later on, until the war in 1914, there were stormy episodes in London and elsewhere. The division of opinion cut across party lines; for example, the Conservative Balfour was pro-woman, and the Liberal Asquith was against. The success of the movement came at the end of the war in 1918.

My political opinions were, and are, on the Liberal

side, as against the Conservatives. I am now writing in terms of English party divisions. The Liberal Party has now (1941) practically vanished; and in England my vote would be given for the moderate side of the Labour Party. However at present there are no "parties" in England.

During our residence at Grantchester, I did a considerable amount of political speaking in Grantchester and in the country villages of the district. The meetings were in the parish schoolrooms, during the evening. It was exciting work, as the whole village attended and expressed itself vigorously. English villages have no use for regular party agents. They require local residents to address them. I always found that a party agent was a nuisance. Rotten eggs and oranges were effective party weapons, and I have often been covered by them. But they were indications of vigour, rather than of bad feeling. Our worst experience was at a meeting in the Guildhall at Cambridge, addressed by Keir Hardie who was then the leading member of the new Labour Party. My wife and I were on the platform, sitting behind him, and there was a riotous undergraduate audience. The result was that any rotten oranges that missed Keir Hardie had a good chance of hitting one of us. When we lived in London my activities were wholly educational.

My philosophic writings started in London, at the latter end of the war. The London Aristotelian Society was a pleasant centre of discussion, and close friendships were formed.

During the year 1924, at the age of sixty-three, I received the honour of an invitation to join the Faculty of Harvard University in the Philosophy Department. I became Professor Emeritus at the close of the session 1936-1937. It is impossible to express too strongly the encouragement and help that has been rendered to me by the University authorities, my colleagues on the Faculty, students, and friends. My wife and I have been overwhelmed with kindness. The shortcomings of my published work, which of course are many, are due to myself alone. I venture upon one remark which applies

to all philosophic work:—Philosophy is an attempt to
express the infinity of the universe in terms of the limitations of language.

It is out of the question to deal with Harvard and its
many influences at the end of a chapter. Nor is such a
topic quite relevant to the purpose of this book. To-day
in America, there is a zeal for knowledge which is reminiscent of the great periods of Greece and the Renaissance. But above all, there is in all sections of the population a warm-hearted kindness which is unsurpassed in
any large social system.

Memories

I

A WAY OF LIFE is something more than the shifting relations of bits of matter in space and in time. Life depends upon such external facts. The all-important æsthetic arises out of them, and is deflected by them. But, in abstraction from the atmosphere of feeling, one behaviour pattern is as good as another; and they are all equally uninteresting. The chief value of memories of infancy and young childhood is that with unconscious naïveté they convey the tonality of the society amid which that childhood was passed. The two generations immediately preceding the present time are so near and so far. We can almost hear the rustle of their clothes as they passed away in the shades. The tones of their voices, their ways of approach, linger. And yet the generation on the younger side of fifty knows so little of them. The blatant emphasis of current literature has done its worst in distortion. Memories shed a quiet light upon ways of feeling which in literature become distorted for the necessities of a story, or of a comparison.

In the autumn of 1864 a small boy three years old was in Paris. He was, however, unconscious of date, of reason, and of personal age. The very notion of the great world of tremendous happenings was absent from his mind. He enjoyed as matter of course the love and petting from the family of parents, children, nurse, and the bright warm days. But one baffling, elusive memory remained throughout life, a thread connecting the child with the onrush of history.

The scene was a bright day, the nurse sitting on a seat facing a broad road, the child playing, a park with its beauty of trees and flowers and shrubs, a palace from which the road came; and whither the road went the child neither knew nor cared. Along the road a glittering regiment of soldiers marched from the palace, and, passing the seat, vanished into the unknown. That was the whole scene, disconnected from any background of date or place, and yet haunting memory in later years. Throughout boyhood he tried again and again to identify the spot. Each year for two months in the late spring he was living in a London house looking across Green Park towards Buckingham Palace. He knew every seat that faced the roads where companies of Queen Victoria's Guards marched to and fro from the palace. The Queen herself, as she drove past, was a familiar sight—a little figure in black, belonging to the unquestioned order of the universe, but at that time, toward the end of the decade of the eighteen-sixties, too retired to be very popular. But the seat of his dream, with its company of soldiers marching from a palace toward the unknown, remained undiscovered.

Years later, in the summer of 1880, I was again in Paris, with my two elder brothers, one of them a schoolmaster, the other a tutor at Oxford. Again, as at the former time, we were returning from Switzerland. Scenes of infancy were entirely out of our thoughts. We were returning to work—the work of the master at an ancient school, of the tutor at Trinity College, Oxford, and of the freshman at Trinity College, Cambridge. We were young men immersed in the academic life of England. The future, like the dream road from the palace to the unknown, lay before us. But suddenly, as I stood in the gardens of the Tuileries, I found the very place of my dreams. The seat was there; the road was there; and the park was there. The dream that had haunted boyhood was discovered to be a reality held in memory.

The vision of the child had caught a glimpse of the pageant of history, and again the second vision gave the tragic interpretation. The palace now stood a ruin, with its charred walls. The Emperor, Napoleon III, had died an exile in England. The road led to Sedan, and the gallant regiments of the French Empire had marched to their doom. The final act of the Napoleonic drama, for which during eighty years Europe was the stage— this final phase, at the glitter of its height and in its downfall, had been flashed upon me in two visions of a seat, a palace, and a road.

At the time of the first vision to the child playing in the garden, secure within his own small world of feelings, human life was exhibiting every diverse phase of horror, enjoyment, and ambition. On September 2, 1864, Atlanta was occupied by the Union forces, and almost immediately Sherman submitted to Grant his plan for his march from Atlanta to the sea—at the very time when the child was playing in the garden. Bismarck was perfecting the policy which brought about the overthrow of Austria within two years. Italy was waiting to seize Rome. The Pope was consolidating his control over the Church, to balance his loss of temporal power. England was nearing the end of the second of its only two long periods of complete security, after the defeat of Louis XIV and after the defeat of Napoleon. Each period was marked by the dominance of a small group of liberal aristocrats.

II

But the history of the world is not focused in any one life. Lincoln had one experience, and his fellow countrymen had each their own experience. The great events that historians speak of influenced more or less directly the lives of all men. But the stuff of human life cannot be wholly construed in terms of historical events; it mainly consists of feelings arising from reactions between small definite groups of persons.

For this reason the generalized history of an epoch sadly misrepresents the real individual feelings of the quiet people in back streets and in country towns. For example, the Victorian epoch in England as seen from our present standpoint entirely misrepresents my memories of the tone of thought of quiet, moderately prosperous people at a time round about the year 1870. I am not talking of agitators, or of people harbouring grievances, but of the ordinary type of leading citizen in a quiet country town. I have already said that the Queen was not popular, and her sanity was doubted. Later she was canonized; but that time was not yet. Also the Prince of Wales, later King Edward the Seventh, was then frankly disliked. The Princess of Wales was beautiful, kindly, and spotless in her conduct; but this only added fuel to the fire as stories passed around. I remember definitely hearing the talk of my elders, that if the Queen died there would have to be a Republic.

From that date the Queen rapidly recovered influence. She became an institution, a legend. Her very individuality, which in the middle period of her reign had annoyed, toward the end became the subject of pride. She was no namby-pamby person who courted popularity. But the Prince of Wales was lucky in the survival of his mother. About twenty years later, in 1890, for a short time he could not appear in public without the escort of the Princess of Wales to subdue the hisses and the ribald shouts. There had been some gambling scandals. But the history of England in the nineteenth century represents a loyal nation gathered lovingly round a spotless throne.

I lived in circles where, if anywhere, loyalty would be found. What really stabilized England was a relatively small group of aristocrats of liberal opinions. These men were highly respected, and had no intention of allowing the country to drift toward any useless experiments. For this reason the desertion of the reforming party by these men over the question of Irish Home Rule, in 1886, was a fatal blow at the old political habits of England.

As to the way in which these men, at the height of their power, managed the Throne, I have been told this story by the son of a cabinet minister who witnessed the incident. During one of Mr. Gladstone's ministries there was a crisis in foreign affairs. The Queen vehemently objected to the policy of the Liberal Cabinet.

For a whole series of cabinet meetings, Mr. Gladstone opened the proceedings by extracting from his dispatch case, with immense solemnity, a letter from the Queen, a new one each time. With growing solemnity, and with all the aid of his magnificent voice, he slowly read Her Majesty's letter. The group of aristocrats who formed the Ministry leaned forward with marked attention to catch every word which emanated from the monarch. The letter always consisted of vehement reproaches to the Ministry for the folly of their conduct. The letter finished, Mr. Gladstone solemnly replaced the document in his dispatch case. The Cabinet then proceeded to business without one word of alluson to the letter, either then or to each other afterward. And the policy of the ministers was never deflected by a hairsbreadth. I doubt if any modern English group of ministers could behave in this way, so inflexibly and with such restraint. But that was the way in which the Whigs ruled the country. In England to-day there is no coherent body of this sort.

This story of Queen Victoria and a group of well-trained politicians is very trivial. It belongs to the frippery of government: how to deal with an awkward incident, of which the importance was more social than political. But the interest is to notice how a score of men with a certain sort of training do in fact deal with such situations. It belongs to the art of preventing minor difficulties from growing into great crises.

It is curious how detached incidents remain in memory. I can vividly remember the old bobbin man who supplied my parents' household with kindling for the coal fires during the first half of the eighteen-seventies. I expect that these bobbins ought to have been called "fagots," but in the villages of East Kent we called them "bobbins." He was a curious old man, completely without education and earning a scanty living. He was dressed in corduroys, of an antiquity defying any exact estimate of date. He cut the scrub undergrowth in the woods near Canterbury, about seventeen miles away from us. He then chopped the wood into the required lengths, and tied the sticks up into parcels—each parcel, or bobbin, being about the amount required to light a fire. He came through the village about once a fortnight, or once in three weeks, with a large cart piled high with bobbins. As he passed, he called out, "Bobbins! Bobbins!" in a curious, harshly rhythmical voice which stays in memory after more than half a century.

The horse was even more decrepit than the man—an old, worn-out cart horse. The pace of the procession was about one and three-quarter miles an hour. They—the man walking beside the horse—plodded along, unresting and untiring, so near their end and yet seemingly timeless and eternal. He, his horse, Queen Victoria, and her cabinet ministers, all belong to the essential stuff of English History. So does my father, the thoroughly countrified vicar of the parish, as I can now see him half a century ago chatting to the old bobbin man. They were on very friendly terms. Unfortunately only one fragment of their conversation survives. It was the old bobbin man who said: "There are some as goes rootling and tearing about. But, Lor' bless you, sir, I gets to Saturday night as soon as any of 'em." That is an authentic bit of village speech, nigh sixty years ago, and the speakers have all passed into their final Saturday night, together with their whole world of ways of life.

III

While on the topic of life in a country vicarage, another visual memory flashes upon me: there is an Archbishop of Canterbury, tall, commanding, stately. He is in a genial mood, with his back to the bright fire in the ample hall of the old vicarage house. He is laughing heartily as my father tells him of the theology of the leading parishioner, who found great comfort in the doctrine of eternal damnation. That incident is also sixty years since. The Archbishop and the leading parishioner must be added to the group of those who make up the stuff of English History.

That Archbishop remains in my memory as one of the few great men whom I have met. I mean men with outstanding governing force conjoined with capacious

intellect. I do not think that he was subtle; but there was
no doubt about him. Archbishop Tait ought to have
been a prime minister. Fate made him Archbishop of
Canterbury. I have always been grateful for my glimpse
of him during half-a-dozen years, and for the family
tradition of him during a longer period. To have seen
Tait was worth shelves of volumes of mediæval history.
He magnificently closed the line of great ecclesiastics who
organized the intimate cultural life of England, round
monasteries, village churches, dioceses, cathedrals, par-
ishes—in New England called "townships"—parish meetings, schools, colleges, universities. The line stretches
from Augustine of Canterbury, through Theodore of
Tarsus, Lanfranc, Anselm, Becket, Warham, Cranmer,
Parker, Laud, Sancroft, Tillotson, Tait. The national
activities that cluster round the archbishops as representative leaders are as much worth dwelling on as those
that centre round kings and parliaments.

Tait really closed the line in the sense in which I am
thinking. All these men from Augustine to Tait energeti-
cally acted on the policy that the Church was the national organ to foster the intimate, ultimate values
which enter into human life. For the earlier men, the
Church was more than that; but at least it was that.
They refused to conceive the Church as merely one
party within the nation, or merely as one factor within
civilization. For them the Church was the nation rising
to the height of its civilization. They were men with
vision—wide, subtle, magnificent. They failed. Tait was
the last Archbishop who effectively sustained the policy.
Since his time, English ecclesiastical policy has been directed to organizing the Anglican Church as a special
group within the nation.

But the failure of the earlier set of men was a magnificent one. Their policy prevailed for twelve hundred
years. It civilized Europe. Country after country has discarded it as an archaic obstruction. Even to-day, Spain
and Mexico are engaged in casting it away. The interest
of men like Warham, Parker, Tillotson, Tait, is that
they rescued the final stage of the mediæval vision of

civilization from the reproach of decrepit reaction. Its end in Spain at the present moment is that of a backward-looking system, divorced from modern realities. Its supporters in Spain are mediæval, blind and deaf to the modern world. But Tait, Tillotson, and Warham, each in his day, were forward-looking men. They took the inherited notion of cultural organization, and tried to give it a new life in terms of the modern world. They failed. Tait was the last of the line. Since his time, smaller men have drifted along with limited aims. Their aims are quite sensible, granting their belief. But they completely fail to stir the blood of those who seek for a vision of civilization in this world.

Perhaps men like Warham, Tillotson, and Tait had gone back behind Christianity to the ideals of Pericles. But to-day, when we are blindly groping for some coherent ordering of civilization, we can spare some sympathy for the men who in England tried to give new life to the old vision which for twelve hundred years had served Europe so well. In England the death of the old ideal had a nobility worthy of its services during its long life.

IV

To return to my theme of memories, we left the Archbishop standing on the vicarage hearthrug, laughing at the silly old gentleman who consoled himself with the thought of the eternal torture of his neighbours.

I can remember the old gentleman well. He was not at all cruel, but simply, incredibly silly. The Archbishop also knew him well, and that is why the religious aspect did not, at the moment, strike him.

Another picture of the old gentleman rises before me. He was taking the chair at a penny reading in the parish schoolroom, which in the evenings acted as an entertainment hall. A penny reading was a series of readings of extracts from good literature—or, at least, what was supposed to be good literature. For the humorous and pathetic pieces Dickens was the favourite author; and among the works of Dickens *Pickwick* was the chief favourite for the comic relief. A certain amount of romantic poetry also was a necessary part—usually Sir Walter Scott. One man did all the reading, someone to whom the parish looked for light and leading in literary matters. For example, a clergyman from a neighbouring parish, or a doctor, or a lawyer; in fact, someone whom the villagers would like to look at for an hour and a half. The entertainment cost a penny, as the name implies. The proceeds about paid for the cost of the gas and of the caretaker. The entertainer was repaid by a supper at the vicarage and a vote of thanks proposed by my father, who usually took the chair. Also I forgot to mention two or three songs, solos, with piano accompaniment, which came between the selections read, and gave the reader a rest. We only rarely rose to a violin solo.

These penny readings spread to every village in southern England at that time. I know nothing about the North of England so far as concerns the details of its life. In the South we were fully occupied with our own village lives. We took no interest in the North of England, which manufactured our linen and woollen clothes; no interest in France, whose cliffs we could see on every fine evening; nor in North America, whose epic of development was the greatest contemporary fact in human history. I am not defending the country folk of East Kent. Facts are stubborn, and it is my present business to state them as I remember.

On the evening in question, my father was the reader at the village penny reading. So the silly old gentleman, as the leading resident, was asked to take the chair. I see him now as though it were yesterday, rising at the close of the meeting, hemming and hawing: "The vicar —he asked me—to thank him—for his great kindness— in so ably entertaining us—and amusing us—this evening." He then had gained his sea legs, and ended quite fluently: "And so, in response to his request, I ask you to join me in thanking him for this magnificent entertainment." On the whole, what he said was the mere truth. But it illustrates how necessary is a decent reserve in the ceremonial of social life.

The penny readings were the first faint signs of a revolution in English culture. Its accomplishment took about fifty years. The England of the eighteenth century and of the main part of the nineteenth century consisted of a highly educated upper class composed of landowners, leaders in business and commerce, and professional men. But the great mass of manual labourers, of artisans, and of the lower end of the traders, were very deficiently educated, if at all. After the middle of the century, and more especially after the first move toward democracy in 1868, the education of the whole nation was seriously initiated. "Let us educate our masters!" exclaimed a leading statesman in a speech in the House of Commons when the plunge had been taken. Of course the movement was slow in getting under way, and still slower in producing any visible effect. But now, looking across fifty or sixty years of conscious recollection, I can see that schools and universities have produced an entirely new type of Englishman, so far as concerns the mass of people.

The standard comments on English education of the earlier period were contained in the Essays of Matthew Arnold. At the time when he wrote they were true enough. But nothing in his Essays applies to the England of to-day. It is still fashionable for superior persons in England to quote him as though his criticisms still applied. But these superior persons are engrossed in reading literature and often have scanty knowledge of the immediate facts around them. One of my most precious memories is that I have, within the space of my lifetime, witnessed the education of England, and the change in English lives that that education has meant.

v

The old bobbin man, as he journeyed with his horse and wagon slowly from the woods near Canterbury to the North Foreland at the tip of Kent, passed through scenes of English History unthinkingly and unknowingly. There still remain in England individuals of his mental grade. But as a type he has vanished from the land. The gap in education between classes has been largely closed. To him the immense story of Canterbury, with its relics of martyrs, heroes, artists, and kings, was as nothing. He jogged along across the meadow marshland with Roman forts on either hand; he passed through the village of Minster, with its magnificent Norman church and its relics of a monastery that once ruled the neighbourhood; he saw the spot where Saint Augustine preached his first sermon; he saw the beach where the Saxons landed; he passed Osengal—that is, the place of bones—perhaps the first English graveyard. But all these things were as nothing to him. He could appreciate neither the past from which he sprang nor the forces of the present which so soon were to sweep away folk like him.

The age of a vast subject population, deaf and dumb to the values belonging to civilization, has gone. Also the old civilizing influence of the Church has passed. It has been replaced by secular schools, colleges, universities, and by the activities of the men and women on their faculties. In the age to come, how will these new agencies compare with the ecclesiastics, the monks, the nuns, and the friars, who brought their phase of civilization to Western Europe?

At the present time, the system of modern universities has reached its triumphant culmination. They cover all civilized lands, and the members of their faculties control knowledge and its sources. The old system also enjoyed its triumph. From the seventh to the thirteenth century, it also decisively altered the mentalities of the surrounding populations. Men could not endow monasteries or build cathedrals quickly enough. Without doubt they hoped to save their souls; but the merits of their gifts would not have been evident unless there had been a general feeling of the services to the surrounding populations performed by these religious foundations. Then, when we pass over another two centuries, and watch the men about the year fifteen hundred, we find an ominous fact. These foundations, which started with such hope and had performed such services, were in full decay. Men like Erasmus could not speak of them without an expression of contempt. Europe endured a hundred years of revolution in order to shake off the system. Men such as Warham, and Tillotson, and Tait struggled for another three centuries to maintain it in a modified form. But they too have failed. With this analogy in mind, we wonder what in a hundred years, or in two hundred years, will be the fate of the modern university system which now is triumphant in its mission of civilization. We should search to remove the seeds of decay. We cannot be more secure now than was the ecclesiastical system at the end of the twelfth century and for a century onward. And it failed.

To my mind our danger is exactly the same as that of the older system. Unless we are careful, we shall conventionalize knowledge. Our literary criticism will suppress initiative. Our historical criticism will conventionalize our ideas of the springs of human conduct. Our scientific systems will suppress all understanding of the ways of the universe which fall outside their abstractions. Our modes of testing ability will exclude all the youth whose ways of thought lie outside our conventions of learning. In such ways the universities, with their scheme of orthodoxies, will stifle the progress of the race, unless by some fortunate stirring of humanity they are in time remodelled or swept away. These are our dangers, as yet only to be seen on the distant horizon, clouds small as the hand of a man.

Those of us who have lived for seventy years, more or less, have seen first the culmination of an epoch, and then its disruption and decay. What is happening when an epoch approaches its culmination? What is happening as it passes toward its decay? Historical writing is cursed with simple characterizations of great events. Historians should study zoology. Naturalists tell us that in the background of our animal natures we harbour the traces of the earlier stages of our animal race. Theologians tell us that we are nerved to effort by the distant vision of ideals, claiming realization. Both sets are right. A daughter of John Addington Symonds, in a novel entitled *A Child of the Alps,* remarks: "Spring is not a season, it is a battleground between summer and winter."

In like manner every active epoch harbours within itself the ideals and the ways of its immediate predecessors. An epoch is a complex fact; and in many of its departments these inherited modes of thought and custom survive, unshaken and dominant. But on the whole the modes of the past are recessive, sinking into an unexpressed background. They are still there, giving a tonality to all that happens, and capable of flaring into a transient outburst when aroused by some touch of genius. Nor is it true that these vanishing ways of thought only appeal to the more backward natures. On the contrary, we find men of capacious intellect and cautious natures endeavouring, in this way and in that way, to adapt the wealth of inheritance to the oncoming fashions of thought. That is how I characterized some of the outstanding Archbishops of Canterbury, from Warham to Tait. Such men disagree in many ways. For example, Tillotson and Tait stand in sharp antagonism to Laud. But they all agree in that they were endeavouring to adapt some generalization of the old ecclesiastical-feudal organization of mankind to the purposes of the dominant rationalistic-individualistic epoch.

We were apt to conceive the Puritans who in the first half of the seventeenth century founded the Commonwealth of Massachusetts as the direct antagonists of these men. But, as we now know, this is a complete mistake. These Puritans were endeavouring to carry over a remodelled ecclesiastical organization as a dominant institution in the new individualistic epoch. In many ways these Puritans are to be classed with Laud, as striving to preserve more of the old world than either Tillotson or Tait.

The true antithesis to all these men is Roger Williams. Curiously enough, this man, who more completely than any other expressed the new individualistic tendencies, seems to stand as an isolated rebel, outside his own times, and yet not fitting into the world of either of the centuries subsequent to his own. He embodied too completely the dominant features of the oncoming world.

In the last seventy years this individualism culminated, retaining as a background the monarchical, aristocratic social ways. These social ways were the recessive retention of the old feudal ecclesiastical system of the Middle Ages. We have watched these ways fading away into the undiscernible inheritance of the past. All that we can now see of them consists of funny little relics here and there—reminding one of the Lion and the Unicorn on the old Boston State House. But with this final triumph of individualism the whole epoch crumbled. New methods of co-ordination are making their appearance, as yet not understood. These principles of organization are based upon economic necessities. That is about all we know of them; the rest is controversy. The older principles of the mediæval system were derived from religious aspirations. Undoubtedly we have lost colour in the foreground by this shift from the ideal to the practical; but the change is more in appearance than in fact. The practical was always there—the hard routine by which the folk of the mediæval times barely sustained life. The difference is that nature controlled them, while we now see our way to the control of nature. That is why the topic of production, distribution, and the organization of labour is now in the foreground.

The other side—the shift in the prominence of the religious motive in social organization—that is too large a topic for the end of a paper.

The Education of an Englishman

WE THINK IN generalities, but we live in detail. To make the past live, we must perceive it in detail in addition to thinking of it in generalities. In this paper I am jotting down recollections of details and generalities of boyhood in an English school, fifty years ago.

Tolstoy has written, as the first sentence of his *Anna Karenina:* "Happy families are all alike; every unhappy family is unhappy in its own way." Thus what is best in English boyhood of that period is identical with what is best in New England experience, of to-day or of that period. But every nation is bad in its own way. We cannot be social reformers all the time. In our off moments we view our peculiar domestic mixture of goods and evils with an affectionate tolerance of their incongruities, which we call "humour." So please remember in reading English literature that the humorous aspects of English life are in general minor symptoms of social defects.

Any account of a phase of national life must throw light on two things: (a) why the nation is as good as it is, and (b) why the nation is as bad as it is. If it be our own country which is in question, the combined complex fact is the country which we love, with its virtues and its defects.

Personal recollections are limited by personal experience. So these pages are not recollections of English education *passim;* but they are typical of one important phase, and apart from knowledge of this phase you cannot understand how England functioned during the latter sixty years of the nineteenth century. The limitations of these recollections can be defined by a reference to Anthony Trollope. His novels refer to the grown-up members of the same society. My recollections refer to the children of the families which he writes about. The fathers of the boys were archdeacons, canons, rectors in the Established Church, or officers in the Army, or small squires in the South-West of England, or lawyers, or doctors. There was a sprinkling of boys from large commercial families.

Most of the moderate capital behind the professional families had come from commerce at no distant date. For us commerce meant trade, banking, ship-owning. Manufactures belonged to the North of England, of which our knowledge was about as vague as it was of the United States. Of course we knew about it, and it was a subject for pride as a national asset, but we did not grasp what it really meant. Anyone who comes from the North of England can reciprocate this indifference of boyhood, from the opposite end.

The school was in Dorsetshire, at Sherborne, a small town of six thousand inhabitants. At that time there were three hundred boys. We were locally termed "The King's Scholars," in allusion to the remodelling of the school in the sixteenth century by King Edward the Sixth. As time was reckoned in that district, this event was still a recent innovation. It was a blot on the scutcheon, introducing a modern vulgarity into what would otherwise have been an unbroken continuity of a thousand years.

Geography is half of character. The soil there is rich, loamy, and gravelly. The climate is formed by warm currents and warm moist winds from the South Atlantic. My own home was in the South-East of England, where we are formed by the polar currents and Siberian winds which come down the North Sea, with interludes of South Atlantic weather from the English Channel. But the interludes in the East were the habitual climate in the West. England is the battleground for these opposed currents, polar on the eastern side, subtropical on the south-western side. Dorsetshire was a rich agricultural district, with apple orchards, and woodlands, and ferns, and rolling grass downs. It did not matter which end of a shrub you put into the ground when planting it; the shrub was bound to grow six feet in the next year. The peasantry had an English dialect of their own, which an Easterner could hardly understand. They were a kindly folk; if a schoolboy on a country walk asked for water, he was given cider and no payment taken.

The town and school had all been founded together by Saint Aldhelm, who died in the year 709, after planting a monastery in that spot. Their importance in the scheme of things has been singularly level from that time on. Perhaps the chief importance came in the eleventh and twelfth centuries, but minor ups and downs hardly count. The most distinguished of the scholars was King Alfred. His connection with the school was mythical, but undoubted. Indeed, vague traditions of the place went back beyond Alfred and beyond Aldhelm to King Arthur, who was said to have held his court on the site of the old British earthworks, amid the neighbouring downs. (Every respectable district in the West of England claims King Arthur.) Certainly when you sat there, on Cadbury Castle, on a warm summer afternoon in the quiet of the dreaming landscape, it seemed eminently probable; and the school song accepted the tradition without question.

So far as sound was concerned, the chief elements were the school bell—a wretched tinkle by which our lives were regulated—and the magnificent bells of the big Abbey Church, which were brought from Tournai by Henry VIII when he returned from the Field of the Cloth of Gold, and given by him to the Abbey. These bells were a great factor in the moulding of the school character, the living voices of past centuries.

This æsthetic background was an essential element in the education, explanatory alike of inertia and of latent idealism. The education cannot be understood unless it is realized that it elucidated an ever-present dream world in our subconscious life.

Some of our classrooms were parts of the old monastery buildings. My own private study in the last two years at school was said to have been the Abbot's cell. The evidence was vague and devoid of documents, but while you lived there it was indubitable. The new school buildings were in the old style, and built of material from the same quarry as that which, centuries earlier, had furnished the stone for the Abbey and the Monastery. This was the Ham Hill quarry. Old Mother Shipton, a prophetess of the early nineteenth century, prophesied that the end of the world would start from Ham Hill. I disbelieve her, because sheer inertia would keep Ham Hill going long after the rest of things had disappeared. To start anything at Ham Hill would constitute a miracle overtaxing credulity.

We had plenty of evidence that things had been going on for a long time. It never entered into anybody's mind to regard six thousand years seriously as the age of mankind—not because we took up with revolutionary ideas, but because our continuity with nature was a patient, visible fact, and had been so since the days of Saint Aldhelm. There were incredible quantities of fossils about; more fossils than stones—or rather, the stones were built out of fossils, welded together.

The boys had thorough country tastes, and knew about the birds, and the ferns, and the foxes, and the gardens. Their fathers rode with the foxhounds, and so did their mothers and sisters. Those who did not hunt planted flowers in their gardens, knew all about the archæology of the neighbourhood, and read Tennyson. Browning would have bothered them. Between whiles, they achieved a good deal of patronage of their social inferiors, with more or less brutality or kindliness, according to breeding and character.

The squire of the district was a very big man, owned half the county, and daily drove his own carriage with four horses—a four-in-hand, as we call it. He was an oldish man then, but he did everything in the grand manner. He and his wife were strict evangelical church people. They must have come under the influence of their neighbour, Lord Shaftesbury, the social reformer. His estates were well managed, with great liberality. This demoralized the neighbourhood, because the "Old Squire" was expected to pay for everything, and did so. He was the chairman of the Board of Governors of the school, and when he died he was succeeded by the Bishop of Salisbury. That sort of alternation had been going on from time immemorial. Nobody thought of it as old habit, or particularly cherished it for that reason; it was just the nature of things—either a Digby or the Bishop; there was no other alternative. Nobody in Sherborne ever did anything explicitly because it was traditional. That is a characteristic of modern progressive societies.

The squire lived in the new castle, a Tudor building of the age of Elizabeth. The old castle was on the other side of the lake in the park. Its Norman keep was blown to pieces by Cromwell's soldiers, after it had been defended against the Parliament by the Countess Digby of that epoch. I do not know why the new castle got itself built half a century before the old castle was knocked down. But after all, the Digbys survived the Puritan soldiers, and so have their political principles of West Country Toryism. To-day the government of England is in the hands of West Country men with an industrial experience—Baldwin and Austen Chamberlain—who are endeavouring to adapt the Digby traditions to modern times. Chamberlain is Birmingham and Worcestershire, and Baldwin is a Shropshire man who has been a large ironmaster. When he was first Prime Minister, some of his workmen made a pilgrimage to Downing Street and held a beanfeast there.

In the old-world woodlands and orchards of the West Country, with its reminiscent landscape, a secret has been whispered down the generations: the secret of governing England in days when kindly sense and tolerance are required to heal its wounds.

The staff of the school, the headmaster and his colleagues, were all strong Liberals, classical scholars, and modernist churchmen. This was in strict accordance with the Rugby tradition, which had been established by Thomas Arnold, a full generation earlier. The Tory squires of the neighbourhood, who governed the school, were conscientious men, and knew how a gentleman should be educated. According to the tradition, which stretched really beyond Arnold, this could only be done efficiently by gentlemen who had read the classics with sufficient zeal to convert them to the principles of Athenian democracy and Roman tyrannicide.

We were taught a good deal of history, very thoroughly so far as it went. But it was characteristically limited according to the prejudices shared equally by the Liberal schoolmasters, the Tory parents, and their children who were the scholars. Our reading was closely limited to those periods of history which, if we might trust our national pride, were closely analogous to our own. We did not want to explain the origin of anything. We wanted to read about people like ourselves, and to imbibe their ideals. When the Bible said, "All these things happened unto them for ensamples," we did not need a higher critic to tell us what was meant or how it came to be written. It was just how we felt.

For example, in Roman history we stopped short at the death of Julius Cæsar. Freedom was over then. A gentleman could no longer say what he liked in the House of Lords or in the House of Commons—that is to say, in the Roman Senate or to the citizens in the Forum. Strictly speaking, we ought to have stopped when Cæsar crossed the Rubicon; but human nature is always illogical, and we—that is to say, masters and scholars—were urged on by curiosity to see how it ended, and also by secret sympathy with Cæsar, who was very like a great English landed magnate of cultivated mind and of sporting tastes, contesting his county parliamentary constituency, with a good chance of being unseated for bribery and corruption. Pompey was unpopular; he lacked the West Country touch. Cicero needed no explanation—he was the Roman substitute for a Lord Chancellor.

These things were not explained to us: the facts spoke

for themselves. We read Tacitus and enjoyed his epi-
grams, though they were hard to translate into English
terse enough to satisfy our masters, and we were not allowed to use English versions. Tacitus carried our sym-
pathies along with him in his denunciation of a state
of society which had lost all close analogy to the British
Constitution. So we made no study of Imperial Rome;
it lacked political interest.

I am not wandering from my subject. I am endeavouring to explain the direct relevance of a classical education half a century ago to the state of mind of an English schoolboy. The prayer which one of us in turn had
to read daily in the school chapel told us that we were
being trained "to serve God in Church and State," and
we never conceived life in any other terms. The com-
petitive conception of modern industry was entirely absent from our minds; also we were ignorant—comparatively ignorant—of the peculiar problems incident in
such a society. The terms in which the Greeks and the
Romans thought were good enough for us. What had
not been said in Greek on political philosophy had not
been said at all.

The Greeks reigned supreme in our minds. Roman
gladiators, Roman debauchery, Roman grandiosity, the
difficulties of writing Latin prose in the style of Cicero,
the absence of a definite article in the Latin language,
the Roman Emperors, and the Popes of Rome, all contributed to a feeling that Rome lacked any true intimate
affinity with us. Looking backward, I think that our instincts were right. The social tone of Dorsetshire in the
eighteen-seventies was really very different from that of
Rome at any time of its history, despite the analogies
which caught our interest.

But Athens was the ideal city, which for two centuries
had shown the world what life could be. I do not affirm
that our image of Athens was true to the facts. It was
something much better; it was alive. The Athenian navy
and the British navy together ruled the seas of our imaginations. It was not oceans that we thought of, but
narrow seas. Oceans are the discovery of the last half-

century, so far as English schoolboys are concerned, and putting Robinson Crusoe aside as the exception to prove the rule. Our navy has never ruled the oceans. It ruled the seas. It caught its enemies rounding capes, or moored in bays, just as the Greeks did. Cape Trafalgar, Cape St. Vincent, and Aboukir Bay were read into Greek history. In those days, half a century ago, our main fleet was in the Mediterranean just where the Greek fleets sailed; and Russia was to us what Persia was to the Greeks. Scholars may demur to this analogy; but I am talking of schoolboys fifty years ago.

Herodotus and Thucydides, with Xenophon on the Ten Thousand, were the successful authors. We all of us cherished a secret hope of travelling in the East. The East then meant the eastern Mediterranean, including Syria and Egypt. Years ago, two twin brothers—my uncles, as it happens—met by accident in a back street of Damascus, neither knowing that the other was out of England. Happy men! They were travelling in the East.

Archæology and learning were secondary matters then, and, as I strongly suspect, are so now to many English archæologists. It was the flavour of the East that we hungered after, the product of our classical education. To understand what I mean, read Kinglake's *Eothen;* it is short and very amusing. It is redolent of English mentality during the mid-nineteenth century.

The Greek insistence on the golden mean and on the virtue of moderation entered into our philosophy of statesmanship, sometimes reinforcing our natural stupidity, sometimes moderating our national arrogance. We conceived India through our knowledge of the East derived from the Greeks. Thus we took an immense interest in Alexander the Great. We forgot the loss of Greek liberty in the thrilling spectacle of a small European army making its way through a vast Eastern Empire. In Alexander at Issus we saw Clive at Plassey.

Decidedly, half a century ago a classical education had a very real relevance to the future lives of these English boys. Among the boys at that small school from 1870 to 1880 were a future commander-in-chief in India, a future general commanding in the Madras Presidency, a future bishop of all southern India. "To serve God in Church and State" was no idle form of words to set before them.

Our school course was a curious mixture of imaginative appeal and precise, detailed knowledge. We had no interest in foreign languages. It was Latin and Greek that we had to know. They were not foreign languages; they were just Latin and Greek; nothing of importance in the way of ideas could be presented in any other way. Thus we read the New Testament in Greek. At school —except in chapel, which did not count—I never heard of anyone reading it in English. It would suggest an uncultivated religious state of mind. We were very religious, but with that moderation natural to people who take their religion in Greek.

The difficulty as to the Old Testament was surmounted by reading the Septuagint in class on Sunday afternoons, though the lower forms had to descend to the vulgarity of the King James Bible. In this Greek presentation of religion the passion for accurate philology sometimes overcame the religious interest. I remember the headmaster stopping a boy who, when translating into English before the assembled class, reeled off the familiar phrase, "Alas, alas, the glory of Israel hath departed," with "No, no, laddie: The glory of Israel has gone away as a colonist."

A few days ago the head of a Canadian university called on me. He turned out to be from the same school; he went there the term after I left. We called to mind these Septuagint lessons, and agreed that in some way they were among the valuable elements of our school training. The Platonizing Jews of Alexandria are mixed in my mind with monastery buildings in Dorsetshire on warm Sunday afternoons in May. When I try to recall how we thought of the Jews, I think that it is accurately summed up in the statement that we believed them to be inspired, but otherwise unimportant.

We studied some mathematics, very well taught; some science and some French, both very badly taught; also some plays of Shakespeare, which were the worst feature of all. To this day I cannot read *King Lear,* having had the advantage of studying it accurately at school. The failure of the science and of the French was not the fault of the masters. An angel from Heaven could not have persuaded us to take them seriously. Again I am not defending us, but am recording facts.

There was a strict monitorial system. In fact, the discipline out of the classroom depended entirely on the head boys in each house. These boys were chosen merely according to their standing in the intellectual life of the school. If the prefects were also athletic and of high character, the system worked very well; otherwise it worked very badly. In my own schooldays, for about half the time it worked badly and for the other half extremely well. There was some teasing, but no gross bullying. When I was "head of the school," I remember caning a boy before the whole school for stealing. Again I am recording, and not defending. I consulted the head-master privately, and he told me that the alternative was expulsion.

In respect to games we were much more independent than modern English schoolboys or undergraduates at any American university. We had lovely playing fields surrounded by intimate scenery such as, in all the world, only the West of England can provide. We managed the games ourselves, and trained ourselves. We played cricket, and football, and fives, because we enjoyed those games and for no other reason. Efficiency, what crimes are committed in thy name! To-day, throughout English schools, the games are supervised by the younger masters. Fifty years ago at Sherborne no master either played a game or interfered with advice, except by the express invitation of the boy who was captain of the games. We were not efficient; we enjoyed ourselves. Also, perhaps in consequence of that freedom from supervision, we were on the best of terms with the masters, and were always pleased when any of the younger members of the staff accepted our invitation to play, an invitation which was regularly forthcoming on every occasion.

In the particular "house"—that is to say, set of dormitories—where I lived, there were ninety boys and four baths. Again I am recording and not defending. Of course there were washbasins in our bedrooms, the water being put there in jugs. Labour was cheap in those days, and plumbing was barely in its infancy. Fifty years before that time, the boys washed under a pump in the school yard. They also managed to serve God in Church and State, so little are some things affected by modern conveniences.

We rose—nominally—at 6.30 a.m. and were in chapel at 7 a.m., if our state of dress, or undress, enabled us to pass the prefect at the chapel door. If not, we had to write out some lines in Greek. I remember cuffing a big boy over the head because I found him twisting the arm of a small boy; but I apologized afterwards, because I found that the small boy had called his elder "a captain of Barbary apes"; this was unpermissible insolence in the school world.

Altogether we were a happy set of boys, receiving a deplorably narrow education to fit us for the modern world. But I will disclose one private conviction, based upon no confusing research, that, as a training in political imagination, the Harvard School of Politics and of Government cannot hold a candle to the old-fashioned English classical education of half a century ago.

England and the Narrow Seas

I

IN ENGLISH RECORDS of the sixteenth and seventeenth centuries there is a phrase which often recurs—"the Narrow Seas." Historians treat it as a name, and tell us, rightly enough, that it refers to the seas which lie just north and south of the Straits of Dover. But what they do not tell us adequately is how greatly the fate of the world has been affected by the peculiarities of these narrow seas. The marked character of these seas has impressed itself upon the populations on its shores: in England these are the East Kent folk and East Anglians from Essex, Suffolk, Norfolk, and Lincolnshire; and on the continent across the water they are the people of the Low Countries—namely, Holland, Belgium, and the north-western coast of France. There are two characteristics impressed on all these populations, with the possible exception of the French section, which has for its hinterland the Latin influence of France. These characteristics are obstinacy and a tendency to lonely thought. There are some things which cannot be learned from State documents in record offices; and one of these facts, which is thus apt to escape notice, is how the Narrow Seas impressed their character on these coastal populations. The Narrow Seas are the parents of all the free governments in the world—Holland, England, the United States. The Pilgrim Fathers were their offspring.

The Straits of Dover form the southern apex of the small triangle in which the North Sea ends; and they form the north-eastern apex of the triangle where the English Channel narrows down to the twenty miles separating England from the civilized world of Latin influence. On the map it looks the simplest job in the world to sail up the Channel, pass through the Straits, and thence up the estuary of the Thames to London. Alternatively there is the short voyage from Antwerp to London. Philip of Spain saw that. Yet there are only four records of a successful invasion across the Narrow Seas: the Romans, the Saxons, William the Conqueror, and the Dutch William the Third. The list suggests high-class efficiency; and it is all wanted for the task. I always suspect that Julius Cæsar and his Roman successors had colossal luck in getting across and in getting back. A fog and a gale, with a Roman fleet wrecked on the treacherous sunken sands or blown on to some dangerous headland—Beachy Head, or the South Foreland, or the North Foreland—might have left England barbarous for another four hundred years and have altered the history of the world. The chances were heavily against those fair-weather Mediterranean sailors, used to tideless, fogless seas. Perhaps Providence sometimes takes a hand in the game of history.

The Narrow Seas put up almost every form of difficulty known to sailors—tides, fogs, winds, dangerous headlands, sunken shoals. The tides are the foundation of most of the trouble. The North Sea and the Channel act as funnels and concentrate their tides at the Straits. The rise and fall in height is a detail compared to the current, which runs like a race-horse. There are four tides a day, two from the north and two from the south. Their relative strengths depend on the winds. Accordingly in the Narrow Seas, four times a day, there is repeated that contest between the North and the South which makes the history of Europe throughout the ages.

These currents have formed shoals which run northward from the Straits of Dover to the mouth of the Thames. My earliest recollections are entwined with flash lights from the lightships on the Goodwin Sands. We could see them on winter evenings from our nursery windows at the top of the house. Sometimes during a fog the boom of a gun would be heard at slow intervals across the sea. It was a ship ashore on the Goodwin Sands. At other times we saw rockets rise mysteriously from the dark waters. It was the Gull lightship signalling a wreck. Next day we were taken down to the harbour, and there was the lifeboat decked with flags: during the night it had been out and had saved the crew of some vessel slowly sinking in the merciless quicksands.

The navigation of the Narrow Seas is the key to Dutch and English history. There are perils in every direction; there are winds and currents to carry you to them; and there are fogs and blinding storms of sleet to hide all knowledge of your whereabouts. The Dutch and English sailors learned their lesson on the Narrow Seas. The Spanish sailors were used to galleys in the tideless Mediterranean and to huge galleons which ran before the trade winds across the open water of the South Atlantic. When it came to fighting for freedom in the Narrow Seas the oar-driven galleys and the unhandy galleons were helpless. It was no use trusting to oars for large ships in the chops of the Channel: and if you could not sail close to the wind you could say your prayers, for your last moment had come.

As you read a history book, compiled by a learned landsman, it is not so easy to understand why the Armada bolted in terror when it had reached its appointed destination between Antwerp and England. King Philip's strategy must have seemed perfect as he sat in his study in Madrid. Freedom was saved for the world because he had ordered his fleet to halt in a death trap for that type of vessel. Such craft could anchor in the Downs or in Calais Roads, but they could only move thence by running before the gale and making a bolt for it up the North Sea.

In our parish registers for the year 1588, my father's predecessor in the vicarage had written, "To-day buried three sailors from the queene's shippes." I read the entry exactly three hundred years afterward, in the same room in which it had been written. Poor nameless men! I wonder whether they ever knew that they had given their lives for the salvation of English freedom.

Every little harbour along that Kentish coast had, and still has, its lifeboat and its luggers, which, by some mysterious art inbred in the population, keep the seas in all weathers—Deal, Ramsgate, Broadstairs, Kingsgate, Margate, all had these lifeboats, and harbours swarming with luggers and fishing smacks.

The fishermen were decidedly "wet" in the technical American sense of that word. I remember one old man who used to row us children out to bathe from his boat. He was a weather-beaten old fellow, and the philosophy of life which he imparted to our eager ears was that "eating is a beastly habit." We all understood, without explanation, that the great-souled way of life was to sustain it on alcoholic beverages—beer for daily life and brandy for festivals. You may criticize the moral code of these men when you have risked your life in saving others as often as had that old lifeboat's man. He shall not remain nameless: his name was Saxby—"Old Saxby" we called him. In his old age, when we were entrusted to him, he got his livelihood by shrimping and by leasing his rowboat. Old Saxby was more remarkable for obstinacy than for lonely thought. But this sole relic of his conversation proves that even he had elaborated his individual outlook on the universe.

The fishing smacks used to trawl in the neighbourhood, and also go farther afield into the North Sea to the Dogger Bank. About every third fish in the North Sea ends by being eaten either in England or in Holland. If you drop a ring, either in Boston or in London, your chance of seeing it again is very small. But if you will send it to the English Fishery Board, they will tie it to the tail of a fish and let it loose in the North Sea; and every third time you will get your ring back.

During the Russo-Japanese War, England and Russia nearly went to war over the fishing smacks on the Dogger Bank. The Russian fleet going from the Baltic to Japan, where the Japanese sank it in *their* Narrow Seas, crossed the Dogger Bank in the night-time and found it studded with small boats and lights. They concluded that they had fallen into an ambush of Japanese torpedo boats, and accordingly opened fire on the fishing smacks. England was aflame with indignation. But luckily Mr. Balfour, the then Prime Minister, and the Lords of the Admiralty—who in England play the august part of your Supreme Court here—kept their heads. The naval officers said that, if you thought you saw a hostile torpedo boat, you had to shoot first and inquire afterwards— since there was not time for the converse procedure— and Mr. Balfour remembered that the Russians were probably ignorant of the peculiarities of the Narrow Seas. So the Russian fleet was allowed to pass through the British squadron, and sailed on to its appointed doom.

II

The history of the world depends on a lot of little things, apart from which events would have happened differently. London would never have been heard of as a great centre of commerce unless just to the north of the Straits of Dover there had been a magnificent anchorage off Deal. It is called the Downs. In English naval history the Downs loom large. With an east wind a sailing ship bound from London southward cannot tack and get round the capes of the South Foreland and Beachy Head. The Narrow Seas, at their narrowest part, forbid that. So in old days the ships from London anchored in the Downs. They wanted good anchorage there: on the French side lies Cape Gris-Nez, on the English side there is the South Foreland, and a few miles behind, ready to engulf them, lie the Goodwin Sands with the treacherous water rippling over them. It is not healthy to be caught in a gale in that spot without good anchorage. The Downs have lost their importance in these days of steam; but in my boyhood I have seen a hundred sail anchored in the Downs. Such a sight might have been seen for centuries, but now the Downs have disappeared from history.

In 1871, during the Franco-German War, an English squadron anchored in the Downs for months. I remember being taken out to see the battleships. In those days all but one had sails as well as steam power. During the Great War it would have been certain destruction to anchor in the Downs. The haunts of my boyhood in Ramsgate fared badly then: a bomb fell on the house where I was born, another in the garden where I played, and a third blew up a powder magazine on the quay where Old Saxby used to embark us for bathing. I do not think anyone left the town by reason of these little incidents. People repaired their windowpanes and stuck it out with East Kent obstinacy. Certainly my own aunt, who still lives there, never moved her establishment.

But at that game of determination Yorkshire beats us hollow. During the war a general examination of all the school children in Scarborough, a seaside town of Yorkshire, had been arranged by the local authorities to take place from nine to twelve in the morning. At six on that morning three German cruisers appeared and shelled the place for over an hour. It never occurred to the authorities to put off the examination, or to the parents to keep the children from school; nor was the work of the children in any way affected. By the time the examination had begun, a British squadron had turned up, and a North Sea fog had descended to save the Germans; so the townspeople did what they always have done in a fog—they went on with their appointed work.

I wonder if you noticed the names of the little Kentish seaports which I mentioned: Ramsgate, Broadstairs, Kingsgate, and Margate. To a man of Kent—Kentishmen are an inferior brand who live at the west end of the county beyond the River Medway—to a man of Kent these names by their very form all suggest the white chalk cliffs of Old England. These cliffs are perpendicular, with "gaps" or "gates" in them at intervals. Wherever there is a gate there is a small fishing town. I suppose that some early Anglo-Saxon pirate got weary of these endless "gates," and so preferred "stairs" for the Broadstairs "gate."

III

When the Anglo-Saxons grew tired of piracy and took to Christianity and the quiet life, they were bothered by the piratical habits of their unconverted cousins in Scandinavia. So all the old villages and churches are about one or two miles inland. Behind the gaps which lie around the headland of the North Foreland there stands a magnificent group of eleventh century Norman churches—Minster, St. Laurence behind Ramsgate, St. Peter's behind Broadstairs, St. John's behind Margate, and Monkton. If you do not understand something about life in the eleventh century when you have visited these, you are incapable of learning. They all have one very useful characteristic. They could hold all the villagers of those times; and when the massive oak doors were shut and barred, from the top of the square Kentish flint tower, with the aid of a few arrows and stones, you could watch the pirates till they went off with the next tide.

I do not mean to imply that the inhabitants were foolishly peaceable; because they weren't. Modern America has nothing to teach East Kent in the way of bootlegging. We finally gave it up with the advent of free trade in 1848. But during the Napoleonic Wars the whole population, country gentlemen, magistrates, and clergy, took a hand in the trade. In those good old days the Established Church showed a surprising liberality of sentiment. The services at Minster Church had sometimes to be interrupted to enable the congregation to remove the brandy from the church vaults to neighbouring marshes on the rumoured approach of the preventive men. In my father's vicarage garden at St. Peter's there were caves with legends of smugglers attached to them.

In recent years the population has been diluted by the influx of Londoners, rich and poor, seeking health from the bracing sea air which comes straight down from the North Pole over the North Sea. But throughout the nineteenth century the East Kent population was devoted to Church and State and moderate Whig principles. My grandfather was a Whig in 1815 when Whiggism was dangerous; he voted Whig in 1832 when Whiggism was all-powerful; and he voted Whig in his old age when Mr. Gladstone triumphed in the early 1870's. Throughout the nineteenth century in East Kent the clergy were the real leaders of the people; bootlegging at the beginning, social reform in the middle—it was all one to them. They were all sturdy Englishmen, clergy and laity together. At the beginning of the century Mr. Harvey, the vicar of St. Laurence, was highly respected, and very deservedly so, though he shared in the jovial habits of that period and sometimes was taken home in a wheelbarrow like Mr. Pickwick. He was a man of energy, and formed the new parish of Ramsgate, which had outgrown its mother village of St. Laurence. His son, Mr. Richard Harvey, was appointed to the new Ramsgate church; and in the second quarter of the century, till 1860, presided there amid universal respect, exhibiting the reformed manners of the new age. In fact he was even High-Church, and introduced an altar cloth with the sacred monogram which can be read as the Latin capital letters IHS. This aroused some Protestant feeling, which was allayed only by the happy conjecture that the letters stood for Jenkins, Harvey, and Snowden— the surnames of the vicar and his two curates. This is an interesting example of how religious strife can be allayed by the ingenuity of scholarship.

The population was very Protestant, but curiously antagonistic to the Nonconformist minority whose theological principles were identical with its own. About 1830 an old gentleman—Townsend was his name—made a vow that if ever he entered a Nonconformist place of worship he hoped that God would make him stick in the doorway. He took the vow seriously, for, when a respected Nonconformist died, during the funeral service he stood outside the church by way of respect, but did not venture into the doorway. In those days there was no honeyed sentiment about the union of the churches.

Throughout the middle of the century the vicar of St. Laurence was Mr. Sicklemore, a considerable landowner who lived in a small park in the parish. He was the incarnation of "Church and State" sentiment. Even in his own time he represented an England that was fast passing. He had a magnificent voice and always preached in black kid gloves. The sermons expressed his sentiments about things in general, frankly expressed in the vernacular. Here is one of his perorations, which modern America might take to heart:—

"This Sunday morning, as I walked through my village, I saw its very walls defaced by advertisements. It's shocking! 'Pon my honour, it's shocking!"

And with that beautiful sentiment he dismissed the congregation. I can well remember Mr. Sicklemore; and I cannot begin to imagine his sentiments if some enterprising medium should evoke him to a knowledge of the modern world.

I think my father was the last example of these East Kent clergymen who were really homogeneous with their people, and therefore natural leaders on all occasions, secular and religious. The present-day English clergy are excellent men, but they are divorced from the soil. My father could remember the arrival of the first railway engine in Ramsgate, and he died at the end of the century. So he exactly represents the period of transformation. He had all the habits of thought of a man who had always taken the lead, not because he thought about it, but because it was the natural thing to do. He was entirely devoid of any artificial tone of "uplift." In fact he hated it, and expressed his opinion of "cant" with direct Saxon vigour. But in his generation a tenderness of tone had crept in, and he was an example of it. When the Baptist minister of the village was dying, my father was the only minister whom he would see. Despite all the differences between their churches, they were both East Kent men; and when they read the Bible together they understood each other without many words.

In his youth he had ridden with the hounds, and had

a magnificent seat on horseback. He had also played cricket with every club in the neighbourhood. He knew all the farmers and the labourers; and in his later years he had christened a fair percentage of them, after playing cricket or hunting with their fathers in earlier days when they were boys together.

He was an equal mixture of a High-Churchman and a Broad-Churchman. His favourite history was Gibbon's *Decline and Fall.* I do not think that any of Gibbon's chapters shocked him; for his robust common sense told him that the people of East Kent, with whom he was quite content, were really very unlike the early Christians. His favourite character in the Bible was Abraham, who exhibits many features to endear him to the East Kent mentality.

My father was a natural orator, equally at home in the pulpit or at a mass meeting either of townspeople or of countrymen. His church was always crammed with the villagers, and with townspeople who had walked some miles, and with Londoners spending their holidays in the district.

These East Kent clergy of the old school had a simpler view of the relations of a pastor to his flock than that which prevails at present. They viewed with disapproval the growth of the complex parochial machinery which obtains at present throughout England. It was a case of one-man rule. They were simple and direct in their methods, and yet they got at the heart of the people in a way denied to the present generation. As they walked through their villages, or across the country footpaths, they stopped and chatted with every man, woman, or child whom they met. They knew all about them—whether their patch of vegetable garden was good or bad, whether they were sober or whether they drank, what their fathers were like, and how their sons had turned out. They had homely advice and kindly sympathy to give. Above all, they saw to it that every child in the village went to school and had an education according to the lights of those days. They visited the schools, listened to the children, patted them on the head, and made friends with the school-teachers. It was a humanizing, kindly influence, which trusted mainly to the mercy of God to save the souls of men.

IV

This corner of Kent is called the Isle of Thanet. The arm of the sea which separates it from the mainland had just ceased to be navigable when the Tudors came to the throne. Now its old bed forms desolate grass flats surrounded by tidal ditches. This flat marshy country is from four to six miles broad and about twenty miles long. It is protected from the North Sea by a dike in the Dutch fashion. The connection with the Low Countries used to be closer than it is now. England supplied the raw materials for the industrial cities, such as Ghent and Bruges. The sixteenth- and seventeenth-century cottages are all identical with the corresponding cottages in Flanders. Sandwich, once the chief naval dockyard of England, is an old Dutch town, so far as its buildings are concerned. Its importance finally ceased in the seventeenth century when its harbour silted up in consequence of the closing of the sea channel between Thanet and the mainland. If you go there, you will find quiet Dutch streets, a glorious Norman church, and in the old Town Hall contemporary pictures of the sea fights with the Dutch. In the intervals of fighting their Protestant kinsfolk for the sake of trade, they got over some Flemish men "cunning in waterworks," as their records say. But even these engineers were powerless against the tides of the Narrow Seas, which remorselessly rolled up sand till Sandwich joined with Ravenna in Italy to exemplify how puny are the efforts of man to stay the hand of Fate.

The witness of Sandwich, the lonely marsh telling of the lost sea passage, and the wonderful group of Norman churches, and in the far distance to the west the towers of Canterbury Cathedral, all proclaim that we are in the midst of a district where events have happened which shaped England. It is natural that it should be so, for we are at the very focus of the Narrow Seas.

Place yourself at the south-west angle of the finest of all these Norman churches, the church of Minster in Thanet, now some four miles from the apex of a large shallow bay dividing the two capes, the North Foreland, in Thanet, and the South Foreland, near Dover on the mainland in Kent. Parts of the church are older than the Normans: the small tower behind us is mainly Saxon, but some of its masonry is Roman. Inside the church there is an oak chest said to have been brought over with William the Conqueror—the heavy luggage of some Norman knight. This is the spot which best overlooks what in old times was the main gateway into England from the French coast. The marshes at our feet stretch up to Canterbury to the west; on the south their sea-shore looks towards France; and on the north another shore touches the estuary of the Thames. Till near the end of the Middle Ages these marshes formed the sea passage; and the traffic to London passed through it, avoiding the dangerous voyage round the North Foreland.

Roman soldiers guarded forts, Richborough and Reculver—Rutupiæ and Regulbium,—which still exist at either end of it. Reculver retains only the foundations, with twin mediæval towers to mark the desolateness of its present site. Richborough still shows the massive Roman walls round the huge enclosure. Then Thanet was an island, and Minster in Thanet overlooked the seaway near the Richborough end. From that position you can see the spot where Hengist and Horsa landed with the first band of Saxons, and also, one hundred and fifty years later, Saint Augustine—the missionary, not the theologian. The first Saxons and the first Christian missionaries landed in Thanet for the same reason, because both they and the inhabitants of Kent felt safer with an arm of the sea between them.

Till the beginning of the nineteenth century an old oak tree could be pointed out near the church, under which Augustine is said to have first preached Christianity to Ethelbert, the king of Kent. All the sermons to be delivered in New England next Sunday morning are derived from that ancestor which still haunts the sea winds in the churchyard of Minster in Thanet.

Ethelbert died at Reculver more than thirteen hundred years ago, and its modern desolateness seems to stand guardian over those simple remote times when the pagan king became Christian. Across the marshes you can on a clear day see the towers of Canterbury Cathedral. In St. Martin's Church, just above the Cathedral, is the font in which Ethelbert was baptized. Even in Ethelbert's time the building was a restoration; it was an old Roman church put in order for his Christian wife. In the Cathedral you are shown the stone in the pavement on which Becket fell as he was murdered by Reginald Fitzurse and his companion knights who with him had hurried across the Narrow Seas from France. "The traitor will never rise again!" cried his murderers. It was a false boast, often repeated on like occasions. Becket is one of the greatest of those traitors who have "risen again" in English history as immortal patriots, glorious for resistance to brute force by whomsoever wielded, King, Parliament, or People. Opposite to this spot, on the other side of the Cathedral, the armour of the Black Prince hangs, reminiscent of the battle of Crécy. In the Cathedral there is a Brenchley chapel. The modern Brenchleys were agricultural labourers in my father's parish, thus exemplifying the rule that the descendants of the mediæval barons are chiefly to be found among the peasantry.

Finally, coming back to modern times and to our observation post in the Minster churchyard, we could see thence, during the Great War, train after train of ammunition, in endless procession, pass along the little branch railway track which runs through the marshes from Canterbury to Minster, and thence past Richborough to Sandwich. Richborough had awakened from the sleep of centuries. At its feet the mouth of a small stream forms a harbourage in the marsh, guarded from the air by the mist which for a thousand years had arisen each night finally to perform this last service to freedom, and protected from the sea by a devious passage amid sand-banks. In my childhood I have watched a horse sucked down into the quicksands of that bay, the rider barely escaping. This spot again became a gateway from England to France. The English ammunition was transported across the Narrow Seas in barges or on train ferries. A battleship was moored with its guns trained on the bay across which the Romans, the Saxons, and Augustine had sailed.

Once more the scene has relapsed into its age-long quiet; and yet, as you stand and absorb it into your being, it takes its character from haunting memories, and from the solitary cry of a sea-gull sounding like a stray echo from the past.

The small tower of Roman and Saxon masonry in the churchyard of Minster in Thanet, facing the Narrow Seas where the North Sea meets the English Channel, and Plymouth Rock, sheltered by Cape Cod from the Atlantic Ocean, are the two spots which mark the two origins—the English origin and the American origin, separated by a thousand years—of a new type of civilized culture, now becoming dominant wherever lands of temperate climate border upon seas and oceans.

An Appeal to Sanity

IN INTERNATIONAL RELATIONS the world alternates between contrasting phases, resulting from variation of emotion between the phases of low and high tension.

In the low phase, a disturbance in one region due to some specific disorder remains local. It does not arouse emotions elsewhere. In such circumstances international relations take the form of local agreements or of local disputes, sometimes culminating in local wars. Determinate finite questions are in this way settled one by one, without reference to each other.

In the phase of high tension, vivid emotions excite each other, and tend to spread throughout the nations, disturbing every variety of topic.

To-day the world is plunged in this second phase of contagious emotion. Thus, in the survey which constitutes this appeal, no item can be considered separately.

What is the justification of "isolation" on the part of a powerful nation, when evil is turbulent in any part of the world?

The answer is that history discloses habitual disorganization among nations, somewhere or other. War is a throw-back from civilization for victors and vanquished, whatever be the initial objects of these crusades. Even presupposing victory, we must weigh carefully the losses against the gains.

Thus the habitual policy should be "Isolation—Unless . . ."

Each nation is a trustee for the fostering of certain types of civilization within areas for which it is directly responsible. Its supreme duty is there. Thus a nation should remain isolated, *unless* (1) the evils of the world threaten this supreme duty, or (2) these evils can be rectified by an effort which will not indirectly defeat the performance of this special duty.[1]

I

Now as to England. This country is a European island with a world-wide co-ordinating influence of many types. The continental civilization of Europe, and its political organizations, develop with singularly little reference to England. Throughout the last four hundred years the keynote of the English policy in Europe has been *safety*, and otherwise *isolation* (non-intervention)—that is, such isolation as is consistent with safety. The result has been that English policy is mainly directed to the western fringe of Europe. Regarding the interior of Europe the interest of England is indirect, and has been so from the Tudor times onward.

To justify this attitude we must refer to the English "world-wide co-ordinating influence," for which the popular designation is the ambiguous term "Empire."

In Burma and India there are almost 400,000,000 people, sensitive, acute, backward in modern techniques, with innumerable diversities. This population is nearly three times that of the United States. It requires, above all, co-ordination of its ancient civilizations with modern techniques. It requires generations of peace. For England, Central Europe is a remote detail compared to this problem—that is to say, it is a detail if, as Englishmen, we consider our supreme duty. Our Empire isolates us from Europe—safety excepted.

Then there is the Mahometan world, beyond the Empire, but influencing and influenced. It lies around the

[1] This article was first published in March, 1939, on the eve of the Second World War. To-day, after the experience of the last seven years, I see no hope for the future of civilization apart from world unity based on sympathetic compromise within a framework of morality which the United Nations Organization now represents.

route to India and within India. It spreads over North Africa, interwoven with English interests in Egypt, the Sudan, and Upper Nigeria. It touches the Atlantic Ocean.

Finally there are the self-governing Dominions, and other districts only partially autonomous. This confederation requires quiet growth. In varying degrees it is sensitive to the disorders of the world.

Thus English policy should be basically non-European. In England excited intellectuals are focused upon Europe. The mass of the population remembers its intimate relationships across the oceans—parents, children, cousins.

To understand English policy and its vacillations one must realize that intellectuals of every social grade are interested in the old European civilization, and that the masses gaze beyond the oceans. In Cornwall you will find in most cottages pictures of mining districts throughout the world; in Cambridgeshire I have presided over a village meeting aroused to a storm of indignation over some army regulation about service abroad. Our best garden boy emigrated to Canada. In Wiltshire there lived near our summer cottage an old man who had been in India, serving in the ranks. Such people have no direct connection with Central Europe. English policy sways between these two foci of interest, and has done so for centuries: Europe and the world.

In the confused sociological topics which constitute international relations, there are no clear issues. Such premises are either before their times or behind their times, and only rarely with their times. Sometimes they have no contact with temporal events. They are useful as suggestions to enlighten the imagination in its dealings with practical affairs.

English foreign interests at the present moment can be vaguely classified under four headings, so far as immediate dangers are concerned—Central Europe, the Mediterranean, the Jews, and the Mahometans.

Central Europe, in its form up to the year 1938, had its origin in the Versailles Treaty and the League of

Nations. Both these fundamental elements—the Treaty and the League—have suffered incessant violations and repudiations by every group of every opinion. In the negotiations which framed the Treaty, and in the subsequent repudiation of the sanctity of the Treaty, America took the lead—perhaps rightly. Thus even the vague sanctity of international law ceases to apply either to the Treaty or to the League. At the present moment they are historical reminiscences. They impose the minimum of obligation. Obligation, in European foreign policy, arises from the facts of the immediate situation and from duty to the future. Formal law can refer only to situations sufficiently stable.

The main motives generating excitement in Central Europe are (1) nationality, based upon various modes of community—such as language, analogies of physiological character, contiguity; (2) doctrines of social organization—liberal, dictatorial, communistic, capitalistic, religious; (3) economic opportunity. None of these motives is completely evil or completely good. Their moral justification depends on the particular circumstances of each case.

The social system of Central Europe is very unstable from the Baltic to the Black Sea, and throughout the Balkan States. There is no complete solution. We can only hope for something that survives with the minimum suppression of dominant aspirations. The point to notice is that war, even if successful, can only increase the malignant excitement. The remedy is peace, fostering the slow growth of civilized feelings. War may be necessary to guard world civilization. But for Central Europe the effective remedy is *peace*.

In Central Europe, the immediate focus of interest has been Czechoslovakia. This is a composite state created by the Treaty of Versailles. All states are composite in origin. The essential question is the mutual agreement of the various factors. The name "Czechoslovakia" tells only half the tale. The full name should be Czecho-Slovak-Magyar-Ruthenian-Polish-Germania.

Having regard to genius, moral heroism, and tormented suffering, the histories of Czechs, Magyars, and Poles present three poignant tragedies which together constitute the tragedy of Central Europe. From century to century, from generation to generation, uncertain boundaries sway to and fro. By choosing your date you can make any claim for any one of them. Each group was surrounded by populations repugnant to itself—for some reason of religion or habit of life. Bohemia, Poland, Hungary, each in its own way tells a tale of the horror of history, and of the genius of mankind. In other words, *tragedy*.

The Great War immensely strengthened feelings of national unity and desires for national independence. The historical reasons for these feelings in different national groups are not to the point. The essential fact is their existence to-day. As peace approached, President Wilson proclaimed the satisfaction of these aspirations after national consolidation as one of the aims of the war. This objective was unanimously accepted by all concerned.

This clarity was deceptive. The Czech State could be made adequately self-sufficient only by including alien groups, for economic reasons and for purposes of defence. Also within it, as in other states, populations were intermixed. Thus, swayed by a legitimate admiration of the Czechs and by hopes for acquiescence in unification, the treaty makers provided the Czech State with an amplitude of extension over a fringe of diverse groups. There was nothing necessarily wrong in this policy. It might have succeeded, in another century, or in the absence of German, Magyar, and Polish states across the border. The plain essential fact remains that the experiment has not succeeded now. Also the revolt can appeal to the great principle of nationality, proclaimed by President Wilson, and in 1918 accepted by the whole world.

Is a world war to be waged in support of the thesis that this great doctrine does not apply to Germans, or to Poles or to Hungarians? At the time of the Versailles settlement some members of the Labour Party in Eng-

land protested against the inclusion of alien populations
in the Bohemian State. After twenty years some of their
successors are prepared to fight for its maintenance. Up
to a few months ago, the very mention of military armament provoked horrified resistance from the same party.
To-day they clamour for a crusade in Central Europe,
depending for success on the intervention of the Heavenly Powers. It is one lesson of history that these last-
mentioned powers are usually on the side of common
sense. Of course, miracles do happen; but it is unwise
to expect them.

II

Since the World War the recovery of Germany has
mainly taken the form of consolidating the Germans of
Europe into a unified German State. This process has
been in accordance with dominant feelings of the populations concerned. Also these feelings are grounded in
a long historical tradition. Between Waterloo and the
Austro-Prussian war of 1866 there existed a loose confederation with Austria and Prussia as its leading members. From the time of Charlemagne to that of the French
Revolution, a period of almost a thousand years, each
century produced some form of Germanic unity, more
or less. This wavering exhibition of unity is termed, in
history, the Holy Roman Empire. Thus the present unification of Germans into Germany is grounded on traditions of feelings which survive the oscillations of
history. It is a sensible policy to respect it. To have a
world war in opposition to this Pan-German movement
would be madness. The United States would be the first
power to adopt an unfriendly neutrality, when the mass
of its population had been aroused to survey the situation. Its widespread attitude of criticism of its allies is
not ignored by European statesmen.

Other nations, whose attitude is relevant to success,
would be even more unfriendly. In fighting to maintain
frontiers of the Czech State, we should be thwarting the
keenest aspirations of the Poles and Hungarians. Thus
we should have against us three great examples in Eu-

rope of thwarted aspiration after national unity. And what would be our justification? The sanctity of the Treaty of Versailles, and the fact that the Czechs would be more prosperous if their pre-existing frontiers were retained. Expansion for the sake of prosperity can be justified only by the reciprocal acquiescence and prosperity of the populations thus included. War on behalf of the frontiers of Czechoslovakia as determined by the Treaty of Versailles would have the weakest moral justification, and would involve active or passive opposition from states whose support is essential for success.

Is Germany to be allowed to extend her direct power over the whole of Central and Western Europe? The answer is that Germany (or any state) should be forcibly prevented when three conditions are fulfilled:—

(1) When she is violently interfering with the development of other states, without the justification of establishing any principle of social co-ordination, acknowledged as of prime importance;

(2) When the consequences of an attempt at forcible prevention will not be worse than the consequences of acquiescence;

(3) When such an attempt can secure its direct object.

In all human affairs abstract notions apply vaguely— more or less; we must be content with approximation. Also reasons merge into each other. For example, these three conditions overlap, and have no sharp distinction. But they do represent large approximations, which sometimes are adequate justifications for action, either separately or jointly.

It has been argued that condition No. 1 is not satisfied in respect to the Czechoslovakian question. But this conclusion bears upon the status of condition No. 2. For, owing to the fact that Poland and Hungary feel the same grievance—namely, that their minorities were included in the Czech State—it follows that a war waged by Britain and France on behalf of the Czechs would have involved Poland and Hungary in unfriendly neutrality, if not in active opposition. The two great Western democracies could not have chosen a worse test case.

Further, neither France nor Great Britain can directly reach the Czech State, to secure its immediate defence. Also, their war preparations still suffer from reliance on a League of Nations with mythical omnipotence. Thus victory could be achieved only by a long-drawn-out war of attrition. The populations of Europe would suffer years of acute misery. Millions of human beings would be killed. The young, active, and enterprising part of the population would supply most of the casualties. Europe would emerge exhausted with its emotions barbarized, its ideals brutalized. Also, Czechoslovakia would have vanished.

In the preceding argument two factors have been omitted: (1) an estimate of Hitler's action in the face of threats; and (2) Russia in the background. Would Hitler have given way if England and France had threatened war? Hitler bears no analogy to the kings, presidents, and prime ministers who achieve their positions by the normal working of established constitutions. Such people can retreat or resign. They retain a great position and high respect. Such men can estimate the consequences of the future with emotions guided by reason as it surveys situations settled as to their general structure.

For rulers such as Hitler and Mussolini the emotional situation is entirely altered. Their own safety and that of their cause depend upon an atmosphere of inflamed emotion. In this way their power arose; in this way it maintains itself. The alternative for them is a dungeon and a firing squad. Hitler is an enraged mystic; that is to say, he belongs to one species of prophet. He is not primarily thinking of personal safety. He is enjoying the hysteria which is the very life-blood of his cause. What is the sense of saying that such a man in such circumstances, knowing the strength of his opportunity with Poland and Hungary wavering, with his armies and air force ready, with his knowledge of the temporary weakness of England owing to the block to armament persistently maintained by idealists out of touch with reality —what is the sense of believing that Hitler, with these emotions and with this opportunity, would allow himself to be bluffed into inaction? It might have happened so, because miracles are always possible.

But, ought this miracle to happen? We have already seen that, for the settlement of Central Europe, the release of the alien populations of Bohemia from inclusion in its state was the very solution advocated by these idealists at the time of the Versailles Treaty. It is the readjustment most likely to appease Europe. If our policy is the appeasement of inflamed emotions by the removal of causes of irritation, this should be our first step. It is unfortunate that the present crisis was required to bring it about. Such is history in all ages.

How is the preceding argument affected by the existence of Russia?

Russia is more than the eastern fringe of Central Europe and the northwestern fringe of China, with armed forces capable of producing predetermined results beyond these borders. We have omitted the one of most decisive importance for the future of the world—namely, the south-eastern boundary, which touches the whole length of the central portion of the Mahometan world.

But Russia is more than its boundaries, just as America is more than its Atlantic and Pacific seaboards. The *Encyclopædia* states, "[Russia] is thus the largest unbroken political unit in the world and occupies more than one seventh of the land surface of the globe." What is happening within this great territory? At times we learn of the execution of a batch of generals, or of a batch of political officers, or of a batch of industrial technicians. But we hardly know the reasons. We know little of the mental and physical health of the men in command. We know nothing of the emotions seething throughout the vast stretch of its population. Has the ideal of national co-ordination superseded the initial ideal of international revolution? We do not know. We gain little from the reports of men, however able and disinterested, who have lived for a few years in Moscow. There are three thousand miles from the Ural Mountains to Vladivostok, and a thousand miles from the

Polish border to the Ural Mountains. It is difficult to fathom the emotional reactions of a hundred and fifty million people scattered over this vast region.

The country has just passed through the greatest sudden revolution in history. A moronic dynasty and an upper class, brilliant in all respects with the single exception of its complete political failure, have been exterminated. The revolution was horrible, but probably beneficial.

One fact seems as well established as any other, in the doubtful maze of Russian policy: namely, Russian statesmen of all parties have a contempt for the liberal democratic type of state, illustrated by America, France, England, Scandinavia, Holland. They have no use for that mode of organization. Suppose that war had been declared, and that the Russian armies had successfully established themselves in Central Europe, with Bohemia as their base. Russian statesmanship would have been all-powerful in that region. Neither England nor France could send a soldier there. Is it sensible to assume that Russian statesmanship would be satisfied to have secured the nice little Czech State on the liberal lines approved by America? Surely we can wipe that dream out of the picture. Poland, Rumania, Hungary, and Yugoslavia would have been in a turmoil, the ultimate issue completely uncertain. Tens of millions would have died. The Russian state organization may be better than the present German state system, but the issue of a Central European war, with Russia involved, may produce any mode of social settlement devised in Heaven or in Hell, or by the usual collaboration of both. The only certainty would be a ghastly slaughter leading to an unknown future. The whole drama would be very exciting for idealists watching from the safety of distance. The great probability is that initially the Russian war machine would be very ineffective. There would be a long war.

Yet again essential factors in this crisis of world history have been omitted—the Mahometan world, Italy, the Jews.

If war by ill chance should break out now, there seems

little doubt that Italy will join Germany. The effect of this alliance immensely strengthens the preceding arguments. France will be hampered on another frontier. The French fleet and part of the English fleet will be tied in the Mediterranean. Our pressure on Germany in the North Sea and the Baltic will be to that extent diminished. The war will be longer and more destructive. Eighty-five million people in Great Britain and France will be facing a hundred and twenty million in Germany and Italy. It will be a long pull. The issue of wars does not wholly depend on the count of populations. Also there is the good hope that Russia would intervene and redress this balance.

But at what a cost! Years of war in Central Europe, and the whole Mediterranean world a turmoil of disorder.

We must now consider the Mahometan world. Recent discussions on international relations seem to have been conducted by one-eyed men. There is a renaissance in progress stretching throughout the great region of the ancient and mediæval civilizations from Persia to Mesopotamia, throughout Asia Minor, Syria, and Arabia, and reaching to Egypt. In these regions civilization was born, and in various transformations it flourished till it was overwhelmed in mediæval times by hordes from Central Asia. The old populations remain, and to-day there is recovery. Persian, Turkish, and the various Arabian nations have able and sensible rulers. Egypt is well governed. But the populations are as yet naïve politically, liable to spasmodic outbreaks.

In case of war, with Italy, Russia, France, and England involved, there can be little doubt that the whole of this central region of the Mahometan faith will be reduced to turmoil. Peace is required. There are two hundred million Mahometans in the world. Are their interests to be neglected in comparison with the importance of retaining four million Germans, Poles, and Hungarians, against their will, as subjects of the Bohemian State?

III

To-day the most universal problem is the relation of the Jews to the various countries in which they dwell. Our modern progressive civilization owes its origin mainly to the Greeks and the Jews. The progressiveness is the point to be emphasized. China and India long ago attained to types of life with more delicate æsthetic and philosophic appreciations, in some respects, than our Western type. But they reached a level and stayed there. The Greeks and the Jews, in the few centuries before and after the beginning of the Christian Era, intensified an element of progressive activity which was diffused throughout the many peoples in the broad belt from Mesopotamia to Spain. Political stability is not the point. We are considering ideals shaping emotions and thus issuing into conduct. This progressive character must be kept in mind. So far as Greeks and Jews were active, progress was not in a rut, degenerating into conservation.

The Roman Empire was a great creation. But no Roman ever disclosed a new idea in religion, in science, in philosophy, in art, in literature, or even in the law which is called Roman. The sustained habit of progressive activity was the discovery of Greeks and Semites in the marvellous thousand years which precede and include the foundation of Christianity.

The Greeks have vanished. The Jews remain.

The Jews are unpopular in many lands. In this fact there is nothing to arouse surprise. In England, with its tendency to relapse into a rut of tradition, the Scotch people were unpopular throughout the eighteenth century, after their union with England in the year 1707. They were performing for England services analogous to those of the Jews for all the races west of India and Central Asia. English literature in the eighteenth century, so far as thought is concerned, would be in a poor way if Scotch and Irish contributions were withdrawn. What brilliance was contributed to English politics throughout the nineteenth century by Gladstone the

Scot, and Disraeli the Jew! They transgressed the average limitations. Apart from ability, differences are quite enough to create prejudices.

Thus, in approaching the Jewish problem as it exists to-day, we are considering one of the factors operative to sustain the many values of life. The question at issue is not the happiness of a finite group. It is the fate of our civilization.

To-day civilization is in danger by reason of a perversion of doctrine concerning the social character of humanity. The worth of any social system depends on the value experience it promotes among individual human beings. There is no one American value experience other than the many experiences of individual Americans or of other individuals affected by American life. A community life is a mode of eliciting value for the people concerned.

It is true that there is a mystic sense of the co-ordination and eternity of realized value. But we here approach the basic doctrine of religion. To attach that co-ordination of value to a finite social group is a lapse into barbaric polytheism.

Further, each human being is a more complex structure than any social system to which he belongs. Any particular community life touches only part of the nature of each civilized man. If the man be wholly subordinated to the common life, he is dwarfed. His complete nature lies idle, and withers. Communities lack the intricacies of human nature. The beauty of a family is derivative from its members. The family life provides the opportunity; the realization lies in the individuals.

Thus socal life is the provision of opportunity. If that opportunity be conceived as complete subordination to the limitations of one community, human nature is dwarfed. Render unto Cæsar the things that are Cæsar's. But beyond Cæsar there stretches the array of aspirations whose co-ordinating principle is termed God. It is not to be found in any one simple community life, either economic or knit by aim at domination. Even a religious community is inadequate. There always remains *solus cum solo*. We have developed a moral individuality; and in that respect we face the universe—*alone*.

This is the justification of that liberalism, that zeal for freedom, which underlies the American Constitution and other various forms of democratic government.

It is the reason why the "totalitarian" doctrine is hateful. Governments are clumsy things, inadequate to their duties. A wise government makes provision for the interweaving of alternative forms of community life. The most valuable part of legal doctrine is concerned with the relation of the State to this indefinite group of communities within, and around, it. In this way an international element becomes an essential factor in human life.

To-day, by the introduction of modern techniques, the inter-relations of human beings throughout this planet have reached an intimate importance far beyond anything dreamt of in past ages, even in the early lifetime of older people now living. Science is international and requires international relations among its societies. Art, literature, religion, and commerce are international.

In the simple age of mediæval Europe, the clergy and the Jews served the main purposes of interweaving the varieties of life into a unity of progress. And the clergy were the representatives of the interaction of Greek and Jewish mentalities in previous centuries.

For two and a half thousand years, Semites have continuously provided suggestion, novelty, and achievement, whereby the life of Europe never lost the subconscious ideal of progress.

Of course the Jews are not the only factor producing progress in Western life. But their services have been immense. Also, in the long run, no written document or artistic structure can perform this service. For example, it is possible, and almost usual, to construe the Bible, Greek literature, and the American Constitution with all the limitations of their periods of origin. And then these heritages from the past are transformed into barriers to progress instead of its foundation. In asserting this danger, I am merely repeating the Catholic doctrine that a living Church is required to interpret lifeless documents. Many living agencies are required to transform our experience of the world that has been into our ideal of the world that shall be.

It is for this reason that the Jews have been a priceless factor in the advance of European civilization. They belong to each nation, and yet they impart a tinge of internationalism. They are eager in respect to concepts relevant to progress, just where we have forgotten them. They have a slight—ever so slight—difference of reaction to those commandments which disclose ideals of perfection. They constitute one of those factors from which each period of history derives its originality.

To-day we are witnessing a relapse into barbarism. The tendency touches every country. But it is centred in Europe. And in Europe Germany is the main seat of the vicious explosion. The general character is over-emphasis on the notion of nationality, producing the ideal of the totalitarian state. The activity, derivative from this debased notion, is the determination to exterminate international factors which exhibit human nature as greater than any state-system. The Jews are the first example of this refusal to worship the state. But religions, arts, and sciences will come next, until mankind are reduced to mean little creatures subservient to the god-state, embodied in some god-man. The worth of life is at stake.

Two problems of pressing importance are made urgent by the anti-Jewish explosion in Germany. How can the Jews in Germany be saved? How can the Jews from Germany and elsewhere be redistributed throughout the world?

It should be realized at once that war is no solution for either of these perplexing duties. An immediate war would probably lead to the massacre of hundreds of thousands of Jews, together with the slaughter of other millions throughout various nations. Europe may be forced into war by the wild lusts of dictatorial states to achieve domination. It is necessary for the democracies to be armed and watchful. But war cannot solve the Jewish question. However successful the crusade, it will leave eighty million Germans with emotions yet more remote from civilized standards.

It is obvious, therefore, that our first task is to undertake the expense of receiving the Jews, and of enabling them to settle elsewhere after such training as is necessary for their new life.

The final problem is the permanent settlement. There is not one solution. There must be many settlements in diverse regions. In considering such districts we must be careful to judge them in reference to the techniques of the present and the future, and to free our imaginations of pictures derived from a vanished past. This caution especially applies to the large stretch down the East Coast of Africa. Hitherto it has been out of the way and remote. But to-morrow, when airplane traffic has developed, the whole coast line will be intimately connected with Egypt, Palestine, and India. The world is on the eve of a development as important and as revolutionary as that produced by the introduction of railways. Disastrous oversights will be committed by people whose imaginations are fettered to past history.

And yet, in other ways, the converse error of neglecting the lesson of history shows ominous signs of hindering the process of settlement.

The later centuries of Turkish domination in the Mahometan world have been a period of decay in civilization, even before the military power began to ebb. It is doubtful whether the capture of Constantinople was not a greater disaster to Mahometans than to Christians. Probably not, because the Muslim world for three centuries merely shared the common fate of Asia when it came into contact with the progressive techniques of Europe.

To-day the tide has turned. Throughout Asia there is a revival. The lesson is being learned. Eastern Asia— namely, China and Japan—is not relevant to this immediate discussion. Consider Southern Asia from Burma and the Malay Peninsula, across India, upward to Persia, across Asia Minor, Syria, and Arabia, across Egypt, across North Africa, and ending at Gibraltar and Nigeria on the shores of the Atlantic. Consider the populations and their cultural influences and the vast stretch of the surface of the world.

Throughout this region, England, France, and Italy exercise various types of influence. English influence is the most extensive, especially in the numerical count of population. So far as indigenous military force is concerned, the Mahometan world is easily the most widespread and important. Also the Mahometan nations are producing vigorous and able rulers, and the Turks have had one recent genius in Kemal Ataturk.

How is this British imperial influence to be characterized? It varies from district to district, and from continent to continent. It touches the two extremes, from direct military rule in a few fortresses to mere diplomatic friendliness, especially with Mahometan nations. The chief feature is the general absence of direct military compulsion, except so far as it is supplied by the active assistance and the passive support of the populations directly concerned. Throughout the whole of this vast region, with its thousands of miles of territories and its hundreds of millions of inhabitants, the number of British soldiers can hardly exceed one hundred thousand men. Also in Great Britain there is no large reserve of soldiers, only a few tens of thousands. These sparse reserves can be quickly transferred to a few spots by transport across the seas. The British Empire in Asia and parts of North Africa is now a co-ordinating agency, actively supported or passively accepted by the populations concerned. It is performing a service, sometimes well, sometimes in mediocre fashion, sometimes very poorly.

How in past times that Empire arose is not to the point. To-day it is introducing throughout its vast populations those sociological habits and those various co-ordinations which will enable them to resume their ancient functions in the advance of civilization.

This Empire is of enormous advantage to Great Britain, chiefly in two ways. In the first place, it promotes British trade in those regions; in fact, the Empire arose from that activity. In the second place, it provides civilian employment for a large proportion of the educated classes. Almost every such family has members spread throughout this area. The very army officers turn into governmental agents, governmental advisers, governmental administrators.

The final ideal is a large friendly co-operation of the populations concerned, each self-governing. This ideal is already realized by the confederation of British Dominions. It is an ideal of gradual growth; only within this century has it dominated British policy.

IV

Finally, the Hebrew National Settlement in Palestine remains for examination. Religion has been and is now the major source of those ideals which add to life a sense of purpose that is worth-while. Apart from religion, expressed in ways generally intelligible, populations sink into the apathetic task of daily survival, with minor alleviations. Throughout the whole continental region under consideration, Palestine is the ideal centre to which various religious faiths converge.

It was the genius of the Jews, their vividness of grasp of the religious problem, which bestowed on Palestine this commanding position. The three Western faiths, Judaism, Christianity, Mahometanism, point thither. The final dispersal of the Jews took place in A.D. 70, when Romans captured Jerusalem. Thus the Jews as a dominating element in the population have been absent for as long a time as they ever occupied the country. It was the Jewish genius that bestowed its radiance upon Palestine—eighteen hundred years ago!

Thus many claims converge on Palestine—the Jewish claim in virtue of bygone occupation and of living genius, the Mahometan claim in virtue of age-long occupation and vivid association, and the Christian claim. It must also be remembered that at the end of the Great War the British would not have been in command of Palestine except for the Arab revolt against Turkey, with Lawrence of Arabia co-ordinating the Arab princes.

Concurrently with this revolt, there is the Balfour Declaration, promising British assistance in the establishment of a National Jewish Home in Palestine, in a manner consistent with the rights of the existing Arab population. The carrying out of the policy presents a complex problem; but the policy in itself expresses the complexity of the keen interests which converge upon Palestine, claiming recognition. The whole question was referred to the Arab chieftains, and at the Peace Conference obtained their passive acquiescence. It must also be noted that the Arab princes of the surrounding states, and the Egyptian and Turkish governments, have been conspicuously careful in refraining from intrusion.

The records of the Middle Ages, during the brilliant period of Mahometan ascendancy, afford evidence of joint association of Mahometan and Jewish activity in the promotion of civilization. The culmination of the Middle Ages even in Christian lands was largely dependent upon this association. Thomas Aquinas received Aristotle from it; Roger Bacon received the foundations of modern science from it. The commercial system of the Italian seaports was a copy of the activities throughout the preceding Dark Ages, carried on by Syrians and Jews.

The association of Jews with the Mahometan world is one of the great facts of history from which modern civilization is derived. The Jewish settlement in Palestine has been established with success, in respect to its immediate aims. It has been supported with ability and self-sacrifice. The result has made it evident that the country is capable of supporting yet larger numbers.

There is one exception to this satisfactory issue of the experiment. The Arabs in Palestine are dissatisfied—not all the Arabs, but large sections who are in open revolt. This serious state of things is probably in part due to lack of statesmanlike initiative on the part of British officials. Some genius was required and failed to appear; perhaps there was positive inefficiency. The situation has not been rescued by them, nor has it been improved by two committees of inquiry dispatched from England.

There is, however, another side to this question, which may produce disaster. Any fusion of Jewish and Arab interests must be produced by the Jews and Arabs themselves. This primary objective of statesmanship seems to have been largely overlooked by the Jewish controlling agencies. It would not be fair to the mass of emigrants from Central Europe to expect from them any insight into the complications of Syrian life; but the controlling agencies in England and the United States might have been asked to show some grasp of the essential objectives.

Unfortunately in public utterances, whatever may have been done behind the scenes, there has predominated the demand that Great Britain should force upon Palestine an unrestricted Jewish domination. In one instance there was even a suggestion that the Jewish agencies should refuse to attend any conference to which dissentient Arabs were to be admitted.

This attitude, if maintained, is signing the death warrant of the Jewish Home in Palestine—perhaps not today, but in the near future. In the region of large political affairs, the test of success is twofold—namely, survival power and compromise.

The literary interest of historians is captured by transitory brilliance. Survival power is the basic factor for political success.

For Palestine any immediate solution which depends on the persistent military might of Great Britain is bound to fail. Within the next century there is every prospect that in times of crisis England will be unable to transport sufficient troops. She cannot be depended on to exercise continuous military domination along the Syrian coast. She may return; but continuity is unlikely.

Any convulsion within the vast area of British influence may occupy her reserves of military strength, which merely amount to an adequate police force. When this happens, a convulsion in Palestine must go its own way. Also, in that neighbourhood, convulsions do happen. Within this century, Armenians have been massacred, and the Greeks have been driven from Asia Minor, which was their homeland for nigh three thousand years.

Most British statesmen are keenly aware that they are primarily a co-ordinating agency, exercising police control, and seeking political structures with intrinsic survival power. Some English statesmen of vigorous decisiveness forget this rôle; they try to decide and impose. They are the failures in modern English history, much beloved by vivid intellectuals. Cromwell in Ireland is an outstanding example in the past, and Carlyle was an admiring intellectual.

The second element in political success is "compromise." The essence of freedom requires political compromise. A clash of interests arises when the social system concerned involves a divergence of aim; compromise means an endeavour to adjust these differences so that the social life shall offer the largest spread of satisfactions. Political solutions devoid of compromise are failures from the ideal of statesmanship.

The tradition of Jewish life does not include any large experience in the political management of the societies throughout which it is spread. Jewish thought naturally concentrates on specific ideals, conceived in the abstract, devoid of compromise and of the requisites for survival.

This characteristic, combined with the ability of the race, is the reason for the incalculable services of the Jews to civilization. They supplied ideals beyond conventional habits. At the same time it explains the failure of the race throughout its long history to maintain stable political structures. Jewish history, beyond all histories, is composed of tragedies.

Christianity was founded in Jerusalem, proclaiming ideals beyond the customary habits of the world. The Christian Church, which gave Europe its modern civilization, was seated in Rome, where the long habit of imperial rule adjusted ideals to immediate necessities. Christianity gained its genius from Judæa, and its survival power from the Roman Empire. In the result, Christianity was a Jewish creation interfused with Roman stability.

To-day another tragedy is crucifying the Jewish race. The work of rescue is again vivified by a prophetic hope —the ideal of a Jewish National Home in the central region of its history.

There is always a condition attached to the success of any ideal seeking embodiment in historic reality. The condition in this case is the co-operation of the Mahometan world. There is good reason to anticipate success; Jewish co-operation was a factor in the great period of Mahometan brilliance. In the present remodelling of the Mahometan world, Jewish skills give the exact assistance that the populations require: Jewish learning can mould Mahometan learning to assimilate modern knowledge; Palestine is placed exactly at the sensitive point where the Western world touches Mahometan life.

The University of Jerusalem, technological schools, modes of agriculture and of manufacture, should extend their influence throughout the Near East. Also care should be taken to avoid the indiscriminate extension of European legal ideas into a social life to which they are alien. Crude notions of personal ownership, or of state dominance, fail to apply to the subtleties of tribal life. A sensitive response to the real facts of the life around is required. The simplicities of abstract thought must be shunned.

These warnings are commonplace. Unfortunately they are required.

In the adjustment of Jews and Arabs, one-sided bargains are to be dreaded. They spell disaster in the future. The hope of statesmen should be to elicit notions of mutual service and of the interweaving of habits so that the diversity of populations should issue in the fulfilment of the varied subconscious claims on life.

There is a new world waiting to be born, stretched along the eastern shores of the Mediterranean and the western shores of the Indian Ocean. The condition for its life is the fusion of Mahometan and Jewish populations, each with their own skills and their own memories, and their own ideals.

War can protect; it cannot create. Indeed, war adds to the brutality that frustrates creation. The protection of war should be the last resort in the slow progress of mankind towards its far-off ideals.

PART II

PHILOSOPHY

Immortality[1]

PREFACE

IN THIS LECTURE the general concept of Immortality will
be stressed, and the reference to mankind will be a
deduction from wider considerations. It will be presupposed that all entities or factors in the universe are
essentially relevant to each other's existence. A complete
account lies beyond our conscious experience. In what
follows, this doctrine of essential relevance is applied to
the interpretation of those fundamental beliefs concerned with the notion of immortality.

I

There is finitude—unless this were true, infinity would
have no meaning. The contrast of finitude and infinity
arises from the fundamental metaphysical truth that
every entity involves an indefinite array of perspectives,
each perspective expressing a finite characteristic of that
entity. But any one finite perspective does not enable
an entity to shake off its essential connection with totality. The infinite background always remains as the unanalysed reason why that finite perspective of that entity

[1] *Ed. Note:* This second part of Professor Whitehead's "Summary" was originally delivered on April 22, 1941, as the Ingersoll
Lecture at the Harvard Divinity School.

has the special character that it does have. Any analysis of the limited perspective always includes some additional factors of the background. The entity is then experienced in a wider finite perspective, still presupposing the inevitable background which is the universe in its relation to that entity.

For example, consider this lecture hall. We each have an immediate finite experience of it. In order to understand this hall, thus experienced, we widen the analysis of its obvious relations. The hall is part of a building; the building is in Cambridge, Mass.; Cambridge, Mass., is on the surface of the Earth; the Earth is a planet in the solar system; the solar system belongs to a nebula; this nebula belongs to a spatially related system of nebulæ; these nebulæ exhibit a system with a finite temporal existence; they have arisen from antecedent circumstances which we are unable to specify, and will transform into other forms of existence beyond our imagination. Also we have no reason to believe that our present knowledge of these nebulæ represents the facts which are immediately relevant to their own forms of activity. Indeed we have every reason to doubt such a supposition. For the history of human thought in the past is a pitiful tale of a self-satisfaction with a supposed adequacy of knowledge in respect to factors of human existence. We now know that in the past such self-satisfaction was a delusion. Accordingly, when we survey ourselves and our colleagues we have every reason to doubt the adequacy of our knowledge in any particular. Knowledge is a process of exploration. It has some relevance of truth. Also the self-satisfaction has some justification. In a sense, this room has solid walls, resting upon a stationary foundation. Our ancestors thought that this was the whole truth. We know that it embodies a truth important for lawyers and for the University Corporation which manages the property. But it is not a truth relevant beyond such finite restrictions.

To-day, we are discussing the immortality of human beings who make use of this hall. For the purposes of this discussion the limited perspectives of legal systems and of University Corporations are irrelevant.

II

"The Immortality of Man"—What can this phrase mean? Consider the term "Immortality," and endeavour to understand it by reference to its antithesis "Mortality." The two words refer to two aspects of the Universe, aspects which are presupposed in every experience which we enjoy. I will term these aspects "The Two Worlds." They require each other, and together constitute the concrete Universe. Either World considered by itself is an abstraction. For this reason, any adequate description of one World includes characterizations derived from the other, in order to exhibit the concrete Universe in its relation to either of its two aspects. These Worlds are the major examples of perspectives of the Universe. The word "evaluation" expresses the elucidation of one of the abstractions by reference to the other.

III

The World which emphasizes the multiplicity of mortal things is the World of Activity. It is the World of Origination: It is the Creative World. It creates the Present by transforming the Past, and by anticipating the Future. When we emphasize sheer Active Creation, the emphasis is upon the Present—namely, upon "Creation Now," where the reference to transition has been omitted.

And yet Activity loses its meaning when it is reduced to "mere creation now": the absence of Value destroys any possibility of reason. "Creation Now" is a matter-of-fact which is one aspect of the Universe—namely, the fact of immediate origination. The notions of Past and Future are then ghosts within the fact of the Present.

IV

The World which emphasizes Persistence is the World of Value. Value is in its nature timeless and immortal.

Its essence is not rooted in any passing circumstance. The immediacy of some mortal circumstance is only valuable because it shares in the immortality of some value. The value inherent in the Universe has an essential independence of any moment of time; and yet it loses its meaning apart from its necessary reference to the World of passing fact. Value refers to Fact, and Fact refers to Value. [This statement is a direct contradiction to Plato, and to the theological tradition derived from him.]

But no heroic deed, and no unworthy act, depends for its heroism, or disgust, upon the exact second of time at which it occurs, unless such change of time places it in a different sequence of values. The value-judgment points beyond the immediacy of historic fact.

The description of either of the two Worlds involves stages which include characteristics borrowed from the other World. The reason is that these Worlds are abstractions from the Universe; and every abstraction involves reference to the totality of existence. There is no self-contained abstraction.

For this reason Value cannot be considered apart from the Activity which is the primary character of the other World. Value is the general name for the infinity of Values, partly concordant and partly discordant. The essence of these values is their capacity for realization in the World of Action. Such realization involves the exclusion of discordant values. Thus the World of Values must be conceived as active with the adjustment of the potentialities for realization. This activity of internal adjustment is expressed by our moral and æsthetic judgments. Such judgments involve the ultimate notions of "better" and "worse." This internal activity of the World of Value will be termed "Valuation," for the purpose of this discussion. This character of Valuation is one meaning of the term Judgment. Judgment is a process of unification. It involves the necessary relevance of values to each other.

Value is also relevant to the process of realization in the World of Activity. Thus there is a further intrusion of judgment which is here called Evaluation. This term will be used to mean the analysis of particular facts in the World of Activity to determine the values realized and the values excluded. There is no escape from the totality of the Universe, and exclusion is an activity comparable to inclusion. Every fact in the World of Activity has a positive relevance to the whole range of the World of Value. Evaluation refers equally to omissions and admissions.

Evaluation involves a process of modification: the World of Activity is modified by the World of Value. It receives pleasure or disgust from the Evaluations. It receives acceptance or rejection: It receives its perspective of the past, and it receives its purpose for the future. This interconnection of the two Worlds is Evaluation, and it is an activity of modification.

But Evaluation always presupposes abstraction from the sheer immediacy of fact: It involves reference to Valuation.

If you are enjoying a meal, and are conscious of pleasure derived from apple-tart, it is the sort of taste that you enjoy. Of course the tart has to come at the right time. But it is not the moment of clock-time which gives importance; it is the sequence of types of value—for instance, the antecedent nature of the meal, and your initial hunger. Thus you can only express what the meal means to you, in terms of a sequence of timeless valuations.

In this way the process of evaluation exhibits an immortal world of co-ordinated value. Thus the two sides of the Universe are the World of Origination and the World of Value. And the Value is timeless, and yet by its transformation into Evaluation it assumes the function of a modification of events in time. Either World can only be explained by reference to the other World; but this reference does not depend upon words, or other explicit forms of indication. This statement is a summary of the endeavour throughout this chapter to avoid the feeble Platonic doctrine of "imitation" and the feebler modern pragmatic dismissal of "immortality."

V

To sum up this discussion: Origination is creation, whereas Value issues into modification of creative action. Creation aims at Value, whereas Value is saved from the futility of abstraction by its impact upon the process of Creation. But in this fusion, Value preserves its Immortality. In what sense does creative action derive immortality from Value? This is the topic of our lecture.

The notion of Effectiveness cannot be divorced from the understanding of the World of Value. The notion of a purely abstract self-enjoyment of values apart from any reference to effectiveness in action was the fundamental error prevalent in Greek philosophy, an error which was inherited by the hermits of the first Christian centuries, and which is not unknown in the modern world of learning.

The activity of conceptual valuation is in its essence a persuasive force in the development of the Universe. It becomes evil when it aims at an impossible abstraction from the communal activities of action. The two worlds of Value and of Action are bound together in the life of the Universe, so that the immortal factor of Value enters into the active creation of temporal fact.

Evaluation functions actively as incitement and aversion. It is Persuasion, where persuasion includes "incitement towards" and "deterrence from" a manifold of possibility.

Thus the World of Activity is grounded upon the multiplicity of finite Acts, and the World of Value is grounded upon the unity of active co-ordination of the various possibilities of Value. The essential junction of the two Worlds infuses the unity of the co-ordinated values into the multiplicity of the finite acts. The meaning of the acts is found in the values actualized, and the meaning of the valuation is found in the facts which are realizations of their share of value.

Thus each World is futile except in its function of embodying the other.

VI

This fusion involves the fact that either World can only be described in terms of factors which are common to both of them. Such factors have a dual aspect, and each World emphasizes one of the two aspects.

These factors are the famous "Ideas," which it is the glory of Greek thought to have explicitly discovered, and the tragedy of Greek thought to have misconceived in respect to their status in the Universe.

The misconception which has haunted philosophic literature throughout the centuries is the notion of "independent existence." There is no such mode of existence; every entity is only to be understood in terms of the way in which it is interwoven with the rest of the Universe. Unfortunately this fundamental philosophic doctrine has not been applied either to the concept of "God," nor (in the Greek tradition) to the concept of "Ideas." An "Idea" is the entity answering questions which enquire "How?" Such a question seeks the "sort" of occurrence. For example, "How did it happen that the motor car stopped?"; the answer is the occurrence of a "redness of lighting" amid suitable surroundings. Thus the special entry of the Idea "Redness" into the world of fact elucidates the special transition of fact which is the stoppage of the car.

A different functioning of "Redness" is the enjoyment of a glorious sunset. In this example, the realized value is evident. A third case is the intention of an artist to paint a sunset. This is an intention towards realization, which is the basic character of the World of Value. But this intention is itself a realization within the Universe.

Thus each "idea" has two sides; namely, it is a shape of value and a shape of fact. When we enjoy "realized value" we are experiencing the essential junction of the two worlds. But when we emphasize mere fact, or mere possibility we are making an abstraction in thought. When we enjoy fact as the realization of specific value, or possibility as an impulse towards realization, we are then stressing the ultimate character of the Universe. This ultimate character has two sides—one side is the mortal world of transitory fact acquiring the immortality of realized value; and the other side is the timeless world of mere possibility acquiring temporal realization. The bridge between the two is the "Idea" with its two sides.

VII

Thus the topic of "The Immortality of Man" is seen to be a side issue in the wider topic, which is "The Immortality of Realized Value": namely, the temporality of mere fact acquiring the immortality of value.

Our first question must be, Can we find any general character of the World of Fact which expresses its adjustment for the embodiment of Value? The answer to this question is the tendency of the transitory occasions of fact to unite themselves into sequences of Personal Identity. Each such personal sequence involves the capacity of its members to sustain identity of Value. In this way, Value-experience introduces into the transitory World of Fact an imitation of its own essential immortality. There is nothing novel in this suggestion. It is as old as Plato. The systematic thought of ancient writers is now nearly worthless; but their detached insights are priceless. This statement can be referred to as expressing the habits of Plato's thought.

The survival of personal identity within the immediacy of a present occasion is a most remarkable character of the World of Fact. It is a partial negation of its transitory character. It is the introduction of stability by the influence of value. Another aspect of such stability is to be seen in the Scientific Laws of Nature. It is the modern fashion to deny any evidence for the stability of natural law, and at the same time implicitly to take such stability for granted. The outstanding example of such stability is Personal Identity.

Let us consider more closely the character of Personal Identity. A whole sequence of actual occasions, each with its own present immediacy, is such that each occasion embodies in its own being the antecedent members of that sequence with an emphatic experience of the self-identity of the past in the immediacy of the present. This is the realization of personal identity. This varies with the temporal span. For short periods it is so overwhelming that we hardly recognize it. For example, take a many syllabled word, such as "overwhelming" which was employed in the previous sentence: of course the person who said "over" was identical with the person who said "ing." But there was a fraction of a second between the two occasions. And yet the speaker enjoyed his self-identity during the pronunciation of the word, and the listeners never doubted the self-identity of the speaker. Also throughout this period of saying that word everyone, including the speaker, was expecting him to finish the sentence in the immediate future beyond the present; and the sentence had commenced in the more distant past.

VIII

This problem of "personal identity" in a changing world of occasions is the key example for understanding the essential fusion of the World of Activity with the World of Value. The immortality of Value has entered into the changefulness which is the essential character of Activity. "Personal identity" is exhibited when the change in the details of fact exhibits an identity of primary character amid secondary changes of value. This identity serves the double rôle of shaping a fact and realizing a specific value.

This preservation of a type of value in a sequence of change is a form of emphasis. A unity of style amid a flux of detail adds to the importance of the various details and illustrates the intrinsic value of that style which elicits such emphasis from the details. The confusion of variety is transformed into the co-ordinated unity of a dominant character. The many become one, and by this miracle achieve a triumph of effectiveness—for good or for evil. This achievement is the essence of art and of moral purpose. The World of Fact would dissolve into the nothingness of confusion apart from its modes of unity derived from its preservation of dominant characters of Value.

IX

Personality is the extreme example of the sustained realization of a type of value. The co-ordination of a social system is the vaguer form. In a short lecture a discussion of social systems must be omitted. The topic stretches from the physical Laws of Nature to the tribes and nations of Human Beings. But one remark must be made—namely, that the more effective social systems involve a large infusion of various sorts of personalities as subordinate elements in their make-up—for example, an animal body, or a society of animals, such as human beings.

Personal Identity is a difficult notion. It is dominant in human experience: the notions of civil law are based upon it. The same man is sent to prison who committed the robbery; and the same materials survive for centuries, and for millions of years. We cannot dismiss Personal Identity without dismissing the whole of human thought as expressed in every language.

X

The whole literature of the European races upon this subject is based upon notions which, within the last hundred years, have been completely discarded. The notion of the fixity of species and genera, and the notion of the unqualified definiteness of their distinction from each other, dominate the literary traditions of Philosophy, Religion, and Science. To-day, these presuppositions of fixity and distinction have explicitly vanished: but in fact they dominate learned literature. Learning preserves the errors of the past, as well as its wisdom. For this reason, dictionaries are public dangers, although they are necessities.

Each single example of personal identity is a special mode of co-ordination of the ideal world into a limited rôle of effectiveness. This maintenance of character is the way in which the finitude of the actual world embraces the infinitude of possibility. In each personality, the large infinitude of possibility is recessive and ineffective; but a perspective of ideal existence enters into the finite actuality. Also this entrance is more or less; there are grades of dominance and grades of recessiveness. The pattern of such grades and the ideal entities which they involve, constitute the character of that persistent fact of personal existence in the World of Activity. The essential co-ordination of values dominates the essential differentiation of facts.

We do not adequately analyse any one personal existence; and still less is there any accuracy in the divisions into species and genera. For practical purposes in the immediate surroundings such divisions are necessary ways of developing thought. But we can give no sufficient definitions of what we mean by "practical purposes" or by "immediate surroundings." The result is that we are confronted with a vague spread of human life, animal life, vegetable life, living cells, and material existences with personal identity devoid of life in the ordinary usage of that word.

XI

The notion of "character," as an essential factor in personal identity, illustrates the truth that the concept of Ideas must be conceived as involving gradations of generality. For example, the character of an animal belongs to a higher grade of ideas than does the special taste of food, enjoyed at some moment of its existence. Also for art, the particular shade of blue in a picture belongs to a lower grade of ideas than does the special æsthetic beauty of the picture as a whole. Each picture is beautiful in its own way, and that beauty can only be reproduced by another picture with the identical design of the identical colours.

Then there are grades of æsthetic beauty, which constitute the ideals of different schools and periods of art.

Thus the variation in the grades of ideas is endless, and it is not to be understood as a single line of increasing generality. This variation may be conceived as a spread involving an infinitude of dimensions. We can only conceive a finite fragment of this spread of grades. But as we choose a single line of advance in such generality, we seem to meet a higher type of value. For example, we enjoy a colour, but the enjoyment of the picture—if it is a good picture—involves a higher grade of value.

One aspect of evil is when a higher grade of adequate intensity is thwarted by the intrusion of a lower grade.

This is why the mere material world suggests to us no concepts of good or evil, because we can discern in it no system of grades of value.

XII

The World of Value contains within itself Evil as well as Good. In this respect the philosophic tradition derived from classical Greek thought is astoundingly superficial. It discloses the emotional attitude of fortunate individuals in a beautiful world. Ancient Hebrew literature emphasizes morality. Palestine was the unhappy battle-ground of opposing civilizations. The outcome in the gifted population was deep moral intuition interwoven with barbaric notions. Hebrew and Hellenic thought are fused together in Christian theology, with considerable loss to the finer insights of both. But Hellenic and Hebrew literature together exhibit a genius of æsthetic and moral revelation upon which any endeavour to understand the functioning of the World of Value must base itself.

Values require each other. The essential character of the World of Value is co-ordination. Its activity consists in the approach to multiplicity by the adjustment of its many potentialities into finite unities, each unity with a group of dominant ideas of value, mutually interwoven, and reducing the infinity of values into a graduated perspective, fading into complete exclusion.

Thus the reality inherent in the World of Value involves the primary experience of the finite perspectives for realization in the essential multiplicity of the World of Activity. But the World of Value emphasizes the essential unity of the many; whereas the World of Fact emphasizes the essential multiplicity in the realization of this unity. Thus the Universe, which embraces both Worlds, exhibits the one as many, and the many as one.

XIII

The main thesis in this lecture is that we naturally simplify the complexity of the Universe by considering it in the guise of two abstractions—namely, the World of multiple Activities and the World of co-ordinated Value. The prime characteristic of one world is change, and of the other world is immortality. But the understanding of the Universe requires that each World exhibits the impress of the other.

For this reason the World of Change develops Enduring Personal Identity as its effective aspect for the realization of value. Apart from some mode of personality there is trivialization of value.

But Realization is an essential factor in the World of Value, to save it from the mere futility of abstract hypothesis. Thus the effective realization of value in the World of Change should find its counterpart in the World of Value:—this means that temporal personality in one world involves immortal personality in the other.

Another way of stating this conclusion is that every factor in the Universe has two aspects for our abstractions of thought. The factor can be considered on its temporal side in the World of Change, and on its immortal side in the World of Value. We have already employed this doctrine in respect to the Platonic Ideas: —they are temporal characterizations, and immortal types of value. [We are using, with some distortion, Plato's doctrine of Imitation.]

XIV

The World of Value exhibits the essential unification of the Universe. Thus while it exhibits the immortal side of the many persons, it also involves the unification of personality. This is the concept of God.

[But it is not the God of the learned tradition of

Christian Theology, nor is it the diffused God of the Hindu Buddhistic tradition. The concept lies somewhere between the two.] He is the intangible fact at the base of finite existence.

In the first place, the World of Value is not the World of Active Creativity. It is the persuasive co-ordination of the essential multiplicity of Creative Action. Thus God, whose existence is founded in Value, is to be conceived as persuasive towards an ideal co-ordination.

Also he is the unification of the multiple personalities received from the Active World. In this way, we conceive the World of Value in the guise of the co-ordination of many personal individualities as factors in the nature of God.

But according to the doctrine here put forward, this is only half the truth. For God in the World of Value is equally a factor in each of the many personal existences in the World of Change. The emphasis upon the divine factor in human nature is of the essence of religious thought.

xv

The discussion of this conclusion leads to the examination of the notions of Life, Consciousness, Memory, and Anticipation.

Consciousness can vary in character. In its essence it requires emphasis on finitude, namely, some recognition of "this" and "that." It may also involve a varying extent of memory, or it may be restricted to the immediacy of the present, devoid of memory, or anticipation. Memory is very variable; and except for a few scraps of experience, the greater part of our feelings are enjoyed and pass. The same statement is true of anticipation.

Our sense-experiences are superficial, and fail to indicate the massive self-enjoyment derived from internal bodily functioning. Indeed human experience can be described as a flood of self-enjoyment, diversified by a trickle of conscious memory and conscious anticipation. The development of literary habits has directed attention to superficial sense-experiences, such as sight and hearing;

the deeper notions of "bowels of compassion," and "loving hearts" are derived from human experience as it functioned three thousand years ago. To-day, they are worn out literary gestures. And yet to-day, a careful doctor will sit down and chat, while he observes the types of bodily experiences of the patient.

When memory and anticipation are completely absent, there is complete conformity to the average influence of the immediate past. There is no conscious confrontation of memory with possibility. Such a situation produces the activity of mere matter. When there is memory, however feeble and short-lived, the average influence of the immediate past, or future, ceases to dominate exclusively. There is then reaction against mere average material domination. Thus the universe is material in proportion to the restriction of memory and anticipation.

According to this account of the World of Activity there is no need to postulate two essentially different types of Active Entities, namely, the purely material entities and the entities alive with various modes of experiencing. The latter type is sufficient to account for the characteristics of that World, when we allow for variety of recessiveness and dominance among the basic factors of experience, namely, consciousness, memory, and anticipation. This conclusion has the advantage of indicating the possibility of the emergence of Life from the lifeless material of this planet—namely, by the gradual emergence of memory and anticipation.

XVI

We now have to consider the constitution of the World of Value arising from its essential embodiment of the World of Fact.

The basic elements in the World of Fact are finite activities; the basic character of the World of Value is its timeless co-ordination of the infinitude of possibility for realization. In the Universe the status of the World of Fact is that of an abstraction requiring, for the completion of its concrete reality, Value and Purpose. Also in the Universe the status of the World of Value is that of an abstraction requiring, for the completion of its concrete reality, the factuality of Finite Activity. We now pass to this second question.

The primary basis of the World of Value is the co-ordination of all possibility for entry into the active World of Fact. Such co-ordination involves Harmony and Frustration, Beauty and Ugliness, Attraction and Aversion. Also there is a measure of fusion in respect to each pair of antitheses—for example, some definite possibility for realization will involve some degree of Harmony and some degree of Frustration, and so on for every other pair of antitheses.

The long tradition of European philosophy and theology has been haunted by two misconceptions. One of these misconceptions is the notion of independent existence. This error has a double origin, one civilized, and the other barbaric. The civilized origin of the notion of independent existence is the tendency of sensitive people, when they experience some factor of value on its noblest side, to feel that they are enjoying some ultimate essence of the Universe, and that therefore its existence must include an absolute independence of all inferior types. It is this final conclusion of the absoluteness of independence to which I am objecting. This error haunted Plato in respect to his Ideas, and more especially in respect to the mathematical Ideas which he so greatly enjoyed.

The second misconception is derived from the earlier types of successful civilized, or half-civilized, social system. The apparatus for preserving unity is stressed. These structures involved despotic government, sometimes better and sometimes worse. As civilization emerged, the social system required such modes of co-ordination.

We have evidence of the Hebrews feeling the inefficiency of casual leadership, and asking for a king—to the disgust of the priests, or at least of the later priests who wrote up the story.

Thus an unconscious presupposition was diffused that a successful social system required despotism. This no-

tion was based on the barbaric fact, that violence was
the primary mode of sustaining large-scale social existence. This belief is not yet extinct. We can see the
emergence of civilized concepts in Greek and Hebrew
social systems, and in the emphasis of the Roman Empire upon the development of a legal system, which was
partially self-sustaining. The Roman legions were mainly
stationed on the borders of the Empire.

But in later Europe the great example of the rise of
civilized notions was set by the monasteries in the early
middle ages. Institutions, such as Cluny in its prime,
upheld the ideal of social systems devoid of violence,
and yet maintaining a large effectiveness. Unfortunately
all human edifices require repair and reconstruction;
but our immense debt to mediæval monasteries should
not be obscured by their need of reform at the end of
that epoch. The clever men of the eighteenth century
expressed in words ideals enacted centuries earlier. In the
modern world the activities of Cluny have been reproduced by the work of convents in regions such as Brit-
tany and New England, but rarely in places where
religion is associated with wealth.

Sociological analysis at the present moment is con-
centrated upon these essential factors which presented
the easiest field. Such a factor was the economic motive;
it would be unfair to ascribe this limited outlook to
Adam Smith, although it certainly dominated his followers in the later generations. Then Idealism was in
the background: the abolition of slavery was its final
effort. The primary example, in the civilization of Europe after the fall of the Western branch of the Roman
Empire, was afforded by the Christian monasteries in
their early period.

XVII

The conclusion of this discussion is twofold. One side
is that the ascription of mere happiness, and of arbitrary power to the nature of God is a profanation. This
nature conceived as the unification derived from the
World of Value is founded on ideals of perfection, moral

and æsthetic. It receives into its unity the scattered effectiveness of realized activities, transformed by the supremacy of its own ideals. The result is Tragedy, Sympathy, and the Happiness evoked by actualized Heroism.

Of course we are unable to conceive the experience of the Supreme Unity of Existence. But these are the human terms in which we can glimpse the origin of that drive towards limited ideals of perfection which haunts the Universe. This immortality of the World of Action, derived from its transformation in God's nature is beyond our imagination to conceive. The various attempts at description are often shocking and profane. What does haunt our imagination is that the immediate facts of present action pass into permanent significance for the Universe. The insistent notion of Right and Wrong, Achievement and Failure, depends upon this background. Otherwise every activity is merely a passing whiff of insignificance.

XVIII

The final topic remaining for discussion opens a large question. So far, this lecture has proceeded in the form of dogmatic statement. What is the evidence to which it appeals?

The only answer is the reaction of our own nature to the general aspect of life in the Universe.

This answer involves complete disagreement with a widespread tradition of philosophic thought. This erroneous tradition presupposes independent existences; and this presupposition involves the possibility of an adequate description of a finite fact. The result is the presupposition of adequate separate premises from which argument can proceed.

For example, much philosophic thought is based upon the faked adequacy of some account of various modes of human experience. Thence we reach some simple conclusion as to the essential character of human knowledge, and of its essential limitation. Namely, we know what we cannot know.

Understand that I am not denying the importance of the analysis of experience: far from it. The progress of human thought is derived from the progressive enlightenment produced thereby. What I am objecting to is the absurd trust in the adequacy of our knowledge. The self-confidence of learned people is the comic tragedy of civilization.

There is not a sentence which adequately states its own meaning. There is always a background of presupposition which defies analysis by reason of its infinitude.

Let us take the simplest case; for example, the sentence, "One and one make two."

Obviously this sentence omits a necessary limitation. For one thing and itself make one thing. So we ought to say, "One thing and another thing make two things." This must mean that the togetherness of one thing with another thing issues in a group of two things.

At this stage all sorts of difficulties arise. There must be the proper sort of things in the proper sort of togetherness. The togetherness of a spark and gunpowder produces an explosion, which is very unlike two things. Thus we should say, "The proper sort of togetherness of one thing and another thing produces the sort of group which we call *two things*." Common sense at once tells you what is meant. But unfortunately there is no adequate analysis of common sense, because it involves our relation to the infinity of the Universe.

Also there is another difficulty. When anything is placed in another situation, it changes. Every hostess takes account of this truth when she invites suitable guests to a party; and every cook presupposes it as she proceeds to cook the dinner. Of course, the statement, "One and one make two" assumes that the changes in the shift of circumstance are unimportant. But it is impossible for us to analyse this notion of "unimportant change." We have to rely upon common sense.

In fact, there is not a sentence, or a word, with a meaning which is independent of the circumstances under which it is uttered. The essence of unscholarly thought consists in a neglect of this truth. Also it is equally the essence of common sense to neglect these differences of background when they are irrelevant to the immediate purpose. My point is that we cannot rely upon any adequate explicit analysis.

XIX

The conclusion is that Logic, conceived as an adequate analysis of the advance of thought, is a fake. It is a superb instrument, but it requires a background of common sense.

To take another example: Consider the "exact" statements of the various schools of Christian Theology. If the leaders of any ecclesiastical organization at present existing were transported back to the sixteenth century, and stated their full beliefs, historical and doctrinal, either in Geneva or in Spain, then Calvin, or the Inquisitors, would have been profoundly shocked, and would have acted according to their habits in such cases. Perhaps, after some explanation, both Calvin and the Inquisitors would have had the sense to shift the emphasis of their own beliefs. That is another question which does not concern us.

My point is that the final outlook of Philosophic thought cannot be based upon the exact statements which form the basis of special sciences.

The exactness is a fake.

Mathematics and the Good

I

ABOUT TWO THOUSAND three hundred years ago a famous lecture was delivered. The audience was distinguished: among others it included Aristotle and Xenophon. The topic of the lecture was The Notion of The Good. The lecturer was competent: he was Plato.

The lecture was a failure, so far as concerned the elucidation of its professed topic; for the lecturer mainly devoted himself to Mathematics. Since Plato with his immediate circle of disciples, the Notion of The Good has disengaged itself from mathematics. Also in modern times eminent Platonic scholars with a few exceptions successfully conceal their interest in mathematics. Plato, throughout his life, maintained his sense of the importance of mathematical thought in relation to the search for the ideal. In one of his latest writings he terms such ignorance "swinish." That is how he would characterize the bulk of Platonic scholars of the last century. The epithet is his, not mine.

But undoubtedly his lecture was a failure; for he did not succeed in making evident to future generations his intuition of mathematics as elucidating the notion of The Good. Many mathematicians have been good men— for example, Pascal and Newton. Also many philosophers have been mathematicians. But the peculiar associations of mathematics and The Good remains an undeveloped topic, since its first introduction by Plato. There have been researches into the topic conceived as an interesting characteristic of Plato's mind. But the doctrine, conceived as a basic truth of philosophy, faded from active thought after the first immediate Platonic epoch. Throughout the various ages of European civilization, moral philosophy and mathematics have been assigned to separate departments of university life.

It is the purpose of the present essay to investigate this topic in the light of our modern knowledge. The progress of thought and the expansion of language now make comparatively easy some slight elucidation of ideas which Plato could only express with obscure sentences and misleading myths. You will understand, however, that I am not writing on Plato. My topic is the connection between modern mathematics and the notion of The Good. No reference to any detailed mathematical theorems will be essentially involved. We shall be considering the general nature of the science which is now in process of development. This is a philosophic investigation. Many mathematicians know their details but are ignorant of any philosophic characterization of their science.

II

Within the period of sixty or seventy years preceding the present time, the progressive civilization of the European races has undergone one of the most profound changes in human history. The whole world has been affected; but the origination of the revolution is seated in the races of western Europe and Northern America. It is a change of point of view. Scientific thought had developed with a uniform trend for four centuries, namely, throughout the sixteenth, seventeenth, eighteenth, and nineteenth centuries. In the seventeenth century, Galileo, Descartes, Newton, and Leibniz elaborated the set of concepts, mathematical and physical, within which the whole movement was confined. The culmination may be placed in the decade from 1870 to 1880. At that time Helmholtz, Pasteur, Darwin, and Clerk-Maxwell were developing their discoveries. It was a triumph which produced the death of the period. The change affects every department of thought. In this chapter I emphasize chiefly the shift in the scope of mathematical knowledge. Many of the discoveries which were effective in producing this revolution were made a century earlier than the decade which is here chosen as the final culmination. But the wide realization of their joint effect took place in the fifty years subsequent to 1880. May I add, as an aside, that in addition to its main topic of mathematics and The Good, this chapter is also designed to illustrate how thought develops from epoch to epoch, with its slow half-disclosures? Apart from such knowledge you cannot understand either Plato, or any other philosopher.

III

In order to understand the change, let us conceive the development of an intellectual life which initiated its growth about the year 1870, at the age of about nine or ten years. The whole story reads like a modern version of a Platonic dialogue—for example, the *Theaetetus* or the *Parmenides*. At the commencement of his intellectual life the child would have known the multiplication table up to twelve-times-twelve. Addition, subtraction, multiplication, and division had been mastered. Simple fractions were familiar notions. The decimal notation for fractions was added in the next two or three years. In this way, the whole basis of arithmetic was soon mastered by the young pupil.

In the same period Geometry and Algebra were introduced. In Geometry, the notions of points, lines, planes, and other surfaces are fundamental. The procedure is to introduce some complex pattern of these entities defined by certain relationships between its parts and then to investigate what other relationships in that pattern are implicitly involved in these assumptions. For example, a right-angled triangle is introduced. It is then proved that—assuming Euclidean Geometry—the square on the hypotenuse is equal to the sum of the squares on the other sides.

This example is interesting. For a child can easily look on a figure of a right-angled triangle—as drawn on the black-board by his teacher—without the notion of the squares on the various sides arising in his consciousness. In other words, a defined pattern—such as a right-angled triangle—does not disclose its various intricacies to immediate consciousness.

This curious limitation of conscious understanding is the fundamental fact of epistemology. The child knew what his teacher was talking about, namely, the right-angled triangle quite evidently suggested on the board by the thick chalk lines. And yet the child did not know the infinitude of properties which were implicitly involved.

The primary factors in the boy's concept of a right-angled triangle—as he looked at the black-board—were points, lines, straightness of lines, angles, right-angles. No one of these notions has any meaning apart from the reference to the all-enveloping space. A point has definite position in space, but does not (as then explained) share in any spatial extension. Lines and straight lines have position and also do share in spatial relations between straight lines. Thus no one of the notions involved in the concept of a right-angled triangle has any meaning apart from reference to the spatial system involved.

IV

At that date, apart from a small selection even among eminent mathematicians, it was presupposed that there was only one coherent analysis of the notion of space; in other words, that any two people talking about space must refer to the same system of relations, provided that you expressed a full analysis of every ramification of their meanings. The aim of mathematics, according to their belief, and according to Plato's belief, and according to Euclid's belief, was the adequate expression of this unique, coherent notion of spatiality. We now know that this notion, which had triumphed for about two thousand four hundred years as the necessary foundation for any physical science, was a mistake. It was a glorious mistake: for apart from the simplification thus introduced into the foundations of thought, our modern physical science would have had no agreed simplification of presuppositions by means of which it could express itself.

Thus, the error promoted the advance of learning up to the close of the nineteenth century. At the close of that period, it obstructed the proper expression of scientific ideas. Luckily the mathematicians—at least some of them—had got ahead of the sober thoughts of sensible men of science, and had invented all sorts of fantastic variations from orthodox geometry. At the turn of the centuries, that is, between 1890 and 1910, it was discovered that these variant types of geometry were of essential importance for the expression of our modern scientific knowledge.

From the faint beginnings of geometry, in Egypt and Mesopotamia, up to the present is a stretch of time extending for almost four thousand years. Throughout the whole period this error of a unique geometry has prevailed. Our notions of to-day have a history of about one hundred to a hundred and fifty years. We enjoy the pleasurable satisfaction that "Now we know."

We shall never understand the history of exact scientific knowledge unless we examine the relation of this feeling "Now we know" to the types of learning prevalent in each epoch. In some shape or other it is always present among the dominant group who are preserving and promoting civilized learning. It is a misapplication of that sense of success which is essential for the maintenance of any enterprise. Can this misapplication be characterized? We may complete the phrase "Now we know" by an adverb. We can mean "Now we know—*in part*"; or we can mean "Now we know—*completely*." The distinction between the two phrases marks the difference between Plato and Aristotle, so far as their influence on future generations is concerned. The notion of the complete self-sufficiency of any item of finite knowledge is the fundamental error of dogmatism. Every such item derives its truth, and its very meaning, from its unanalysed relevance to the background which is the unbounded Universe. Not even the simplest notion of arithmetic escapes this inescapable condition for existence. Every scrap of our knowledge derives its meaning from the fact that we are factors in the universe, and are dependent on the universe for every detail of our experience. The thorough sceptic is a dogmatist. He enjoys the delusion of complete futility. Wherever there is the sense of self-sufficient completion, there is the germ of vicious dogmatism. There is no entity which enjoys an isolated self-sufficiency of existence. In other words, finitude is not self-supporting.

The summarized conclusion of this discussion is that geometry, as studied through the ages, is one chapter of the doctrine of Pattern; and that Pattern as known to finite discrimination, is a partial disclosure with an essential relevance to the background of the Universe. Also the term "Geometry" refers to a genus of patterns; and this genus includes a variety of species.

v

We now turn to the discussion of Number, considered as a fundamental mathematical notion. This section can be shortened, because many relevant deflections have already been expressed in the previous examination of Geometry.

The doctrine of number from the Greek period onwards has always included queer little contradictions which thoughtful people disregarded. In the last quarter of the nineteenth century, a more thorough examination of the whole subject, initiated by Georg Cantor and Frege in Germany and Austria, and by Peano and Pieri in Italy, and in England by students of symbolic logic, disclosed a number of awkward questions. Finally Bertrand Russell produced a peculiarly glaring self-contradiction in the current reasoning. I well remember that he explained it to Frege in a private letter. Frege's answer commenced with the exclamation, "Alas, arithmetic totters!"

Frege was correct: Arithmetic tottered and still totters. But Bertrand Russell was equal to the occasion. We were then in the midst of writing a book entitled, *Principia Mathematica*. Russell introduced the notion of "types" of entities. According to that doctrine, the notion of number should only be applied to a group of entities of the same type. Thus the number "three" as applied to entities of one type has a different meaning to the number "three" as applied to entities of another type. For example, if we are considering two different types, there are two different meanings of the number "three."

Russell was perfectly correct. By confining numerical reasoning within one type, all the difficulties are avoided. He had discovered a rule of safety. But unfortunately this rule cannot be expressed apart from the presupposition that the notion of number applies beyond the limitations of the rule. For the number "three" in each type, itself belongs to different types. Also each type is itself of a distinct type from other types. Thus, according to the rule, the conception of two different types is nonsense, and the conception of two different meanings of the number three is nonsense. It follows that our only way of understanding the rule is nonsense. It follows that the rule must be limited to the notion of a rule of safety, and that the complete explanation of number awaits an understanding of the relevance of the notion of the varieties of multiplicity to the infinitude of things. Even in arithmetic you cannot get rid of a subconscious reference to the unbounded universe. You are abstracting details from a totality, and are imposing limitations on your abstraction. Remember that a refusal to think does not imply the non-existence of entities for thought. Our conscious thought is an abstraction of entities from the background of existence. Thought is one form of emphasis.

VI

Finally in this survey of mathematical notions we come to Algebra. Who invented Algebra? It was invented "in Arabia" or "in India," you all want to tell me. In one sense that is true—namely, the useful symbolism for the algebraic ideas started in one or other, or in both, of those countries. But there is a further question, which, I am sure, would have interested Plato if he had known about Algebra. Who invented the fundamental idea which is thus symbolized?

What is the fundamental notion at the base of Algebra? It is the notion of "*Any* example of a given sort, in abstraction from some particular exemplification of the example or of the sort."

VII

The first animal on this Earth, who even for a moment entertained this notion, was the first rational creature. You can observe animals choosing between *this thing* or *that thing*. But animal intelligence requires concrete exemplification. Human intelligence can conceive of a type of things in abstraction from exemplification. The most obvious disclosures of this characteristic of humanity are mathematical concepts and ideals of the Good—ideals which stretch beyond any immediate realization.

Any practical experience of exactness of realization is denied to mankind: Whereas mathematics, and ideals of perfection, are concerned with exactness. It is the difference between practice and theory. All theory demands exact notions, somewhere or other, however concealed. In practice exactness vanishes: the sole problem is, "Does it Work?" But the aim of practice can only be defined by the use of theory; so the question "Does it Work?" is a reference to theory. Also the importance of theory resides in its reference to practice. The vagueness of practice is energized by the clarity of ideal experience.

No one has ever observed in practice any exact mathematical notion. Consider the child as he learnt his geometry. He never observed an exact point or an exact line, or exact straightness, or an exact circle. Such things were unrealized ideals in the child's mind. So much will be conceded by the man of practical good sense. But when we pass to arithmetic he stalls. You can hear him saying—perhaps you are saying it yourselves—"I can see one chair, two chairs, three chairs, four chairs, and five chairs, and I can observe that two chairs and three chairs when assembled together form a group of five chairs." In this way, our sensible friend has observed exactly exemplifications of arithmetical notions and of an arithmetic theorem.

Now the question is—Has he observed exactly, or, Has he had exact notions elicited in his conceptual experience? In what sense did he observe exactly one chair? He observed a vague differentiation of the general context of his visual experience. But suppose we pin him down to one billionth of an inch. Where does the chair end and the rest of things begin? Which atom belongs to the chair, and which atom belongs to surrounding space? The chair is perpetually gaining and losing atoms. It is not exactly differentiated from its surroundings, nor is it exactly self-identical as time slips by. Again, consider the chair during long periods. It gradually changes, even throughout its solid wooden parts. At the end of a million years in a cave, it becomes fragile, and dissolves at a touch. A slow, imperceptible change is always in progress.

Remember that the human concepts of one inch in length, and of one second of time, as being reasonable basic quantities, are purely relevant to human life. Further, the modern discoveries of physicists and astronomers have disclosed to us the relevance of minute, and of immense, happenings. Our exact conceptual experience is a mode of emphasis. It vivifies the ideals which invigorate the real happenings. It adds the perception of worth and beauty to the mere transition of sense-experience. It is by reason of the conceptual stimulus that the sunset displays the glory of the sky. By this statement, it is not meant that a feeble train of explicit thoughts works the miracle. It is the transformation of the real experience into its ideal limit. Our existence is invigorated by conceptual ideals, transforming vague perceptions.

We cannot understand the flux which constitutes our human experience unless we realize that it is raised above the futility of infinitude by various successive types of modes of emphasis which generate the active energy of a finite assemblage. The superstitious awe of infinitude has been the bane of philosophy. The infinite has no properties. All value is the gift of finitude which is the necessary condition for activity. Also activity means the origination of patterns of assemblage, and mathematics is the study of pattern. Here we find the essential clue which relates mathematics to the study of the good, and the study of the bad.

VIII

You will notice that earlier in this essay we have emphasized that there are no self-existent finite entities. The finite essentially refers to an unbounded background. We have now arrived at the converse doctrine, namely, that infinitude in itself is meaningless and valueless. It acquires meaning and value by its embodiment of finite entities. Apart from the finite, the infinite is devoid of meaning and cannot be distinguished from nonentity. The notion of the essential relatedness of all things is the primary step in understanding how finite entities require the unbounded universe, and how the universe acquires meaning and value by reason of its embodiment of the activity of finitude.

Among philosophers, Spinoza emphasized the fundamental infinitude and introduced a subordinate differentiation by finite modes. Also conversely, Leibniz emphasized the necessity of finite monads and based them upon a substratum of Deistic infinitude. Neither of them adequately emphasized the fact that infinitude is mere vacancy apart from its embodiment of finite values, and that finite entities are meaningless apart from their relationship beyond themselves. The notion of "understanding" requires some grasp of how the finitude of the entity in question requires infinity, and also some notion of how infinity requires finitude. This search for such understanding is the definition of philosophy. It is the reason why mathematics, which deals with finite patterns, is related to the notion of the Good and to the notion of the Bad.

The great religions illustrate this doctrine. Buddhism emphasizes the sheer infinity of the divine principle, and thereby its practical influence has been robbed of energetic activity. The followers of the religion have lacked impulse. The doctrinal squabbles of Christianity have been concerned with the characterization of the infinite in terms of finitude. It was impossible to conceive energy in other terms. The very notion of goodness was conceived in terms of active opposition to the powers of evil, and thereby in terms of the limitation of deity. Such limitation was explicitly denied and implicitly accepted.

IX

The history of the science of algebra is the story of the growth of a technique for representation of finite patterns. Algebra is one chapter in the large technique, which is language. But, in the main, language indicates its meanings by means of casual associations as they arise in human history. It is true that language strives to embody some aspects of those meanings in its very structure. A deep sounding word embodies the deep solemnity of grief. In fact, the art of literature, vocal or written, is to adjust the language so that it embodies what it indicates.

But the larger part of what language physically presents is irrelevant to the meaning indicated. The sentence is a sequence of words. But this sequence is, in general, irrelevant to the meaning. For example, "Humpty-Dumpty sat on a wall" involves a sequence which is irrelevant to the meaning. The wall is in no sense subsequent to Humpty-Dumpty. Also the posture of sitting might have been realized simultaneously with the origination of the sitter and the wall. Thus the verbal order has the faintest reference to the idea conveyed. It is true that by exciting expectation, and by delay, the verbal order does work on the emotions of the recipient. But the sort of emotion, thus aroused, depends on the character of the recipient. Algebra reverses the relative importance of the factors in ordinary language. It is essentially a written language, and it endeavours to exemplify in its written structures the patterns which it is its purpose to convey. It may not be always wholly successful in this endeavour. But it does invert the ordinary habits of language. In the usage of Algebra, the pattern of the marks on paper is a particular instance of the pattern to be conveyed to thought.

Also there is an enlargement of the notion of "any." In arithmetic we write "two plus three" equals "three plus two." We are considering two processes of assemblage. The type of assemblage is indicated by the word —or sign—"plus," and its meaning is restricted by the reference to number. The two procedures are asserted to issue in groups with identical number of members. This number is in fact "five"; but it is not mentioned.

Now in algebra, the restriction of thought to particular numbers is avoided. We write "$x + y = y + x$" where x and y are any two numbers. Thus the emphasis on pattern, as distinct from the special entities involved in the pattern, is increased. Thus algebra in its initiation involved an immense advance in the study of pattern. Relationships of diverse patterns, such as that involved in the Binomial Theorem, entered into human thought. Of course, algebra grew slowly. For centuries it was conceived as a mode of asking for the solution of equations. Somewhere in mediæval times, an unfortunate emperor, or other bigwig, together with his court, had to listen to a learned Italian expounding the solution of a cubic equation. Poor men—a lovely Italian afternoon was wasted! They would have yawned if their interest had not been sustained by the sense of magic.

x

At the beginning of the nineteenth century, Algebra was the study of patterns involved in the various ways of assembling numbers, so that each assemblage issued in the indication of one number, conceived as the outcome of that assemblage. The relation of equality between two assemblages meant that both assemblages indicated the same number. But the interest was directed to the two patterns of assemblage, with their identical indications. In this way, certain general characteristics of patterns of number as realized in the evolving universe were identified with characteristics of patterns of marks on two-dimensional surfaces—usually sheets of paper. Such identities of pattern of meaning with pattern of written marks, or sound variation, are a subordinate characteristic of ordinary language, though of some importance in respect to spoken language. But this identity is the major characteristic of algebraic language.

To-day, surveying the first half of the twentieth century, we find an immense extension of algebra. It has been extended beyond the field of number, and applies to a large group of patterns in which number is a subordinate factor. Very often when number is explicitly admitted, its major use is to provide names, as it is employed for the naming of houses. Thus mathematics is now being transformed into the intellectual analysis of types of pattern.

The notion of the importance of pattern is as old as civilization. Every art is founded on the study of pattern. Also the cohesion of social systems depends on the maintenance of patterns of behaviour; and advances in civilization depend on the fortunate modification of such behaviour patterns. Thus the infusion of pattern into natural occurrences, and the stability of such patterns, and the modification of such patterns, is the necessary condition for the realization of the Good.

Mathematics is the most powerful technique for the understanding of pattern, and for the analysis of the relationships of patterns. Here we reach the fundamental justification for the topic of Plato's lecture. Having regard to the immensity of its subject-matter mathematics, even modern mathematics, is a science in its babyhood. If civilization continues to advance, in the next two thousand years the overwhelming novelty in human thought will be the dominance of mathematical understanding.

The essence of this generalized mathematics is the study of the most observable examples of the relevant patterns; and applied mathematics is the transference of this study to other examples of the realization of these patterns.

XI

Pattern is only one factor in our realization of experience, either as immediate value or as stimulus to activity for future value. For example, in a picture, the geometrical pattern may be good, but the relationship of colours may be horrible. Also each individual colour may be poverty-stricken, indeterminate, and feeble. This example elicits the truth that no entity is merely characterized by its individual character, or merely by its relationships. Each entity possesses essentially an individual character, and also is essentially a terminal of relationship, potential or actual. Some of the factors of individual character enter into the relationships, and conversely the relationships enter into the character. In other words, no entity can be considered in abstraction from the universe, and no entity can be divested of its own individuality. The traditional logic overstressed the notion of individual character. The notion of "any" frees us from individual character: but there is no entity which is merely "any." Thus when algebra is applied, factors beyond algebraic thought are relevant to the total situation. Returning to the picture, mere geometry is not the whole tale. Colours are relevant.

In a picture colour (including black and white) may be reduced to a minimum, as in a pen-and-ink sketch. But some differentiation of colour is necessary for the physical presentation of geometrical design. On the other hand, colour may be dominant in some glorious work of art. Again, the drawing may be good, and colour effect may be a failure. The whole topic of Good and Evil arises. And you cannot discuss Good and Evil without some reference to the interweaving of divers patterns of experience. The antecedent situation may demand depth of realization, and a thin pattern may thwart conceptual expectation. There is then the evil of triviality—a sketch in place of a full picture. Again, two patterns eliciting intense experience may thwart each other. There is then the intense evil of active deprivation. This type has three forms: a concept may conflict with a reality, or two realities may conflict, or two concepts may conflict.

There may be other types of evil. But we are concerned with the maladjustment of patterns of experience. The total pattern has inhibited the insistent effect of either of its parts. But this notion is meaningless except as a reference to the background of feeling—namely emotional and analytic experience—within which that total pattern arises. Every abstraction derives its importance from its reference to some background of feeling, which is seeking its unity as one individual complex fact in its immediate present. In itself a pattern is neither good nor bad. But every pattern can only exist in virtue of the doom of realization, actual or conceptual. And this doom consigns the pattern to play its part in an uprush of feeling, which is the awakening of infinitude to finite activity. Such is the nature of existence: it is the acquisition of pattern by feeling, in its emphasis on a finite group of selected particulars which are the entities patterned—for example, the spatial arrangements of colours and sounds. But the particulars concerned are not necessarily purely qualitative. A human being is more than an assortment of colours and sounds. The notion of pattern emphasizes the relativity of existence, namely, how things are connected. But the things thus connected are entities in themselves. Each entity in a pattern enters into other patterns, and retains its own individuality in this variety of existence. The crux of philosophy is to retain the balance between the individuality of existence and the relativity of existence. Also each individual entity in one pattern may be capable of analysis, so as to display itself as the unity of achieved pattern. The point that I am emphasizing is the function of pattern in the production of Good or Evil in the finite unit of feeling which embraces the enjoyment of that pattern. Also the essential characterization of mathematics is the study of pattern in abstraction from the particulars which are patterned.

XII

When Plato in his lecture connected mathematics with the notion of the Good, he was defending—consciously or unconsciously—the traditional ways of thought spread through all races of mankind. The novelty was the method of abstraction which the Greek genius was gradually emphasizing. Mathematics, as studied in his own Academy, was an abstraction of geometrical and numerical characterizations from the concrete facts of Athenian life. Aristotle was dissecting animals, and was analysing political constitutions. He conceived of genera and species. He thus abstracted the logical characters from the full-blooded experience. The new epoch of scientific abstractions was arising.

One danger in the use of this technique is the simpleminded use of Logic, whereby an erroneous proposition is merely discarded. All propositions are erroneous unless they are construed in reference to a background which we experience without any conscious analysis. Every scientific proposition which the great scientists of the mid-nineteenth century entertained, was erroneous in the sense in which it was then construed. Their doctrine of space was wrong: their doctrine of matter was wrong: their doctrines of evidence were wrong. The abiding interest of Plato's Dialogues does not lie in their enunciation of abstract doctrines. They are suffused with the implicit suggestion of the concrete unity of experience, whereby every abstract topic obtains its interest.

XIII

Abstraction involves emphasis, and emphasis vivifies experience, for good, or for evil. All characteristics peculiar to actualities are modes of emphasis whereby finitude vivifies the infinite. In this way Creativity involves the production of value-experience, by the inflow from the infinite into the finite, deriving special character from the details and the totality of the finite pattern.

This is the abstraction involved in the creation of any actuality, with its union of finitude with infinity. But consciousness proceeds to a second order of abstraction whereby finite constituents of the actual thing are abstracted from that thing. This procedure is necessary for finite thought, though it weakens the sense of reality. It is the basis of science. The task of philosophy is to reverse this process and thus to exhibit the fusion of analysis with actuality. It follows that Philosophy is not a science.

Process and Reality

AMERICANS ARE ALWAYS warm-hearted, always apprecia-
tive, always helpful, but they are always shrewd; and
that is what makes for me the continual delight of living in America, and it is why when I meet an American
I always expect to like him, because of that always
delightful mixture of shrewdness and warm-heartedness.

Of course anybody who has any sense who writes on
philosophy knows, or ought to know, that the world is
unfathomable in its complexity and that anything you
put together must be open to criticism—ought to be
open to criticism if it is any good at all. It should be a
platform from which it is worth while to make criticisms.
That is, to be reasonably successful as a philosopher is
to provide a new platform; perhaps not a completely
new platform, but a slight alteration of some older
platform from which it is worth while to make criticisms.
And criticism is the motive power for the advance of
thought. I am fond of pointing out to my pupils that
to be refuted in every century after you have written is
the acme of triumph. I always make that remark in
connection with Zeno. No one has ever touched Zeno
without refuting him, and every century thinks it worth
while to refute him.

May I be a little egotistical at first before going on to
philosophy? I cannot help thinking backwards to a little
less than seven years ago, at the end of August when our
ship with my wife and myself steamed into the port of
Boston. We knew no one resident in Harvard, although
a few weeks before we had met Mr. and Mrs. Osborn

Taylor in London. We steamed in on a wonderful morning at the end of August after we had encountered a summer hurricane the previous night. We felt very small and very wee, wondering what was going to happen to us. And on the quay there stood three of the most welcoming persons, Taylor, Woods, and Edgar Pierce. I remember the heartening feeling of this first expression of the warm and overwhelming kindness that at once laid the foundation of a series of intimate friendships—not taken by a careful selection from this or that remote group, but the intimate friendships with those exact people with whom I have the honour to serve Harvard as one of their colleagues. It seems to have been my good fortune to fall among a set of intimate, stimulating friends, a good fortune which started on the quay with that welcome from Woods, Taylor, and Edgar Pierce—a good fortune that has continued ever since. I do feel that if a man is going to do his best he ought to live in America, because there the treatment of any effort is such that it stimulates everything that is eager in one.

I should like also to be still a little more egotistical. I feel to-day that the mathematical habit of reckoning in quantities is sometimes wrong. I am seventy years old, and I am hanged if I feel it! That is one of the thoughts that come to me.

Another thing I should like to mention to the younger members here who are of course in revolt against the previous age. You have in me a typical example of the Victorian Englishman. I have been struck with the fact that every cause I have in any way voted for in England has finally reached such triumphs as a cause can reach. I have never, never been at final variance with the bulk of my countrymen. I have sometimes and generally voted in a minority for a few years before the cause has triumphed. Most of my votes have been minority votes, but they have always ended in that final majority which settles the question. And thus I deduce that I can have no claim whatsoever to standing above, or beyond, or in any way outside of my age. I am exactly an ordinary example of the general tone of the Victorian Englishman, merely one of a group.

I now pass on to philosophy. I said very little in my book *Process and Reality* about Hegel for a very good reason. You remember that the greater part of my professional life was passed as a mathematician, lecturing and teaching mathematics, and a great deal of the rest has been devoted to the elaboration of symbolic logic. So you will not be surprised when I confess to you that the amount of philosophy I have not read passes all telling, and that as a matter of fact I have never read a page of Hegel. That is not true. I remember when I was staying with Haldane at Cloan I read one page of Hegel. But it is true that I was influenced by Hegel. I was an intimate friend of McTaggart almost from the very first day he came to the University, and saw him for a few minutes almost daily, and I had many a chat with Lord Haldane about his Hegelian point of view, and I have read books about Hegel. But lack of first-hand acquaintance is a very good reason for not endeavouring in print to display any knowledge of Hegel.

But, as I said in my book, I admit a very close affiliation with Bradley, except that I differ from Bradley where Bradley agrees with almost all the philosophers of his school and with Plato, insofar as Plato was a Hegelian. I differ from them where they all agree in their feeling of the illusiveness and relative unreality of the temporal world. Bergson takes the opposite point of view; he holds that the intellect necessarily falsifies the notion of process. There are these two prevalent alternative doctrines respecting the process apparent in the external world: one, which is Bergson's view, is that the intellect in order to report upon experienced intuition must necessarily introduce an apparatus of concepts which falsify the intuition; the other is that process is a somewhat superficial, illusory element in our experience of the eternally real, the essentially permanent. The latter is Bradley's standpoint, if I read him correctly. I think that it is at times Plato's view also. It is exactly on these points that I differ from Bergson on the one side, and from Bradley on the other.

I speak from very thin knowledge; but I rather suspect that I am a little more Aristotelian than either Bergson or Bradley. I would, however, be very sorry to go to the stake for this belief. Aristotle has some very relevant suggestions on the analysis of becoming and process. I feel that there is a gap in his thought, that just as much as becoming wants analysing so does perishing. Philosophers have taken too easily the notion of perishing. There is a trinity of three notions: being, becoming, and perishing. Plato states the question (Plato raises all fundamental questions without answering them) by introducing the notion of that which is always becoming and never real. The world is always becoming, and as it becomes, it passes away and perishes. Now that notion of perishing is covered up as a sort of scandal. Broad even goes so far as to say, in effect, that the past is nothing, *simpliciter*. Again Plato raises the question, when he points out that not-being is a form of being, that whatever you can say about things which are not-being is a way of saying that they have being. He is merely thinking of his forms as including alternative possibilities, when he is making these remarks in the *Sophist*. But like all Plato's remarks it bears thinking about and expanding. We can stretch it to mean that the world as it passes perishes, and that in perishing it yet remains an element in the future beyond itself.

Almost all of *Process and Reality* can be read as an attempt to analyse perishing on the same level as Aristotle's analysis of becoming. The notion of the prehension of the past means that the past is an element which perishes and thereby remains an element in the state beyond, and thus is objectified. That is the whole notion. If you get a general notion of what is meant by perishing, you will have accomplished an apprehension of what you mean by memory and causality, what you mean when you feel that what we are is of infinite importance, because as we perish we are immortal. That is the one key thought around which the whole development of *Process and Reality* is woven, and in many ways I find that I am in complete agreement with Bradley.

I think Bradley gets into a great muddle because he accepts the language which is developed from another point of view. Most of the muddles of philosophy are, I think, due to using a language which is developed from one point of view to express a doctrine based upon entirely alien concepts.

As to my own views of permanence and transience, I think the universe has a side which is mental and permanent. This side is that prime conceptual drive which I call the primordial nature of God. It is Alexander's *nisus* conceived as actual. On the other hand, this permanent actuality passes into and is immanent in the transient side.

Enlarge your view of the final fact which is permanent amid change. In its essence, realization is limitation, exclusion. But this ultimate fact includes in its appetitive vision all possibilities of order, possibilities at once incompatible and unlimited with a fecundity beyond imagination. Finite transience stages this welter of incompatibles in their ordered relevance to the flux of epochs. Thus the process of finite history is essential for the ordering of the basic vision, otherwise mere confusion. The key to metaphysics is this doctrine of mutual immanence, each side lending to the other a factor necessary for its reality. The notion of the one perfection of order, which is (I believe) Plato's doctrine, must go the way of the one possible geometry. The universe is more various, more Hegelian.

Again the attainment of that last perfection of any finite realization depends on freshness. Freshness provides the supreme intimacy of contrast, the new with the old. A type of order arises, develops its variety of possibilities, culminates, and passes into the decay of repetition without freshness. That type of order decays; not into disorder, but by passing into a new type of order.

I certainly think that the universe is running down. It

means that our epoch illustrates one special physical type of order. For example, this absurdly limited number of three dimensions of space is a sign that you have got something characteristic of a special order. We can see the universe passing on to a triviality. All the effects to be derived from our existing type of order are passing away into trivialities. That does not mean that there are not some other types of order of which you and I have not the faintest notion, unless perchance they are to be found in our highest mentality and are unperceived by us in their true relevance to the future. The universe is laying the foundation of a new type, where our present theories of order will appear as trivial. If remembered, they would be remembered or discerned in the future as trivialities, gradually fading into nothingness. This is the only possible doctrine of a universe always driving on to novelty.

Now I have said enough about the philosophy, except that I should like to remark that the modern phases of mathematics or mathematical logic are not modern at all, but arise out of a great past: Grassman, Sir William Hamilton—not the Scotchman who was a bad metaphysician, but the Irishman who wrote good mathematics (when this William Hamilton was ten years old, the Persian ambassador came to Dublin, and this boy was the only available person who could make a public speech in Persian welcoming the ambassador)—Boole, De Morgan, and to go back to the origin of all such efforts, the great Leibniz. I think it is well to cherish that notion of the world's growth of ideas from generation to generation.

John Dewey and His Influence

PHILOSOPHY IS A widespread, ill-defined discipline, performing many services for the upgrowth of humanity. John Dewey is to be classed among those men who have made philosophic thought relevant to the needs of their own day. In the performance of this function he is to be classed with the ancient stoics, with Augustine, with Aquinas, with Francis Bacon, with Descartes, with Locke, with Auguste Comte. The fame of these men is not primarily based on the special doctrines which are the subsequent delight of scholars. As the result of their activities the social systems of their times received an impulse of enlightenment, enabling them more fully to achieve such high purposes as were then possible.

By reason of the Stoics, the subsequent legal tradition of the Western World was securely founded in the Roman Empire; by reason of Augustine Western Christianity faced the Dark Ages with a stabilized intellectual tradition; Aquinas modernized, for the culmination of the Middle Ages, this ideal of a co-ordination of intimate sources of action, of feeling, and of understanding. The impress on modern life due to Bacon, Descartes, Locke, and Comte, is too recent to need even a sentence of reminder.

John Dewey has performed analogous services for American civilization. He has disclosed great ideas relevant to the functioning of the social system. The magnitude of this achievement is to be estimated by reference to the future. For many generations the North American

Continent will be the living centre of human civilization. Thought and action will derive from it, and refer to it.

We are living in the midst of the period subject to Dewey's influence. For this reason there is difficulty in defining it. We cannot observe it from the outside in contrast to other periods also viewed in the same way. But knowledge outruns verbal analysis. John Dewey is the typical effective American thinker; and he is the chief intellectual force providing that environment with coherent purpose. Also wherever the influence of Dewey is explicitly felt, his personality is remembered with gratitude and affection.

II

The human race consists of a small group of animals which for a small time has barely differentiated itself from the mass of animal life on a small planet circling round a small sun. The Universe is vast. Nothing is more curious than the self-satisfied dogmatism with which mankind at each period of its history cherishes the delusion of the finality of its existing modes of knowledge. Sceptics and believers are all alike. At this moment scientists and sceptics are the leading dogmatists. Advance in detail is admitted: fundamental novelty is barred. This dogmatic common sense is the death of philosophic adventure. The Universe is vast.

Dewey has never been appalled by the novelty of an idea. But it is characteristic of all established schools of thought to throw themselves into self-defensive attitudes. Refutation has its legitimate place in philosophic discussion: it should never form the final chapter. Human beliefs constitute the evidence as to human experience of the nature of things. Every belief is to be approached with respectful enquiry. The final chapter of philosophy consists in the search for the unexpressed presuppositions which underlie the beliefs of every finite human intellect. In this way philosophy makes its slow advance by the introduction of new ideas, widening vision and adjusting clashes.

The excellence of Dewey's work in the expression of

notions relevant to modern civilization increases the danger of sterilizing thought within the puny limitations of to-day. This danger, which attends the tradition derived from any great philosopher, is augmented by the existing success of modern science. Philosophy should aim at disclosure beyond explicit presuppositions. In this advance Dewey himself has done noble work.

Analysis of Meaning

PHILOSOPHY IN ITS advance must involve obscurity of expression, and novel phrases. The permanent, essential factors governing the nature of things lie in the dim background of our conscious experience—whether it be perceptual or conceptual experience. The variable factors first catch our attention, and we survive by reason of our fortunate adjustment of them. Language has been evolved to express "clearly and distinctly" the accidental aspect of accidental factors. But no factor is wholly accidental. Everything which in any sense is something thereby expresses its dependence on those ultimate principles whereby there are a variety of existences and of types of existences in the connected universe.

Thus the task of philosophy is to penetrate beyond the more obvious accidents to those principles of existence which are presupposed in dim consciousness, as involved in the total meaning of seeming clarity. Philosophy asks the simple question, What is it all about?

In human experience, the philosophic question can receive no final answer. Human knowledge is a process of approximation. In the focus of experience there is comparative clarity. But the discrimination of this clarity leads into the penumbral background. There are always questions left over. The problem is to discriminate exactly what we know vaguely.

The endeavour to make our utmost approximation to analysis of meaning is human philosophy. For a being with complete knowledge, philosophy would take another aspect. He might say, "Knowing everything, I will fix attention on this detail." He will then enjoy the detail in its relation to the discriminated totality.

We enjoy the detail as a weapon for the further discrimination of the penumbral totality. In our experience there is always the dim background from which we derive and to which we return. We are not enjoying a limited dolls' house of clear and distinct things, secluded from all ambiguity. In the darkness beyond there ever looms the vague mass which is the universe begetting us.

The besetting sin of philosophers is that, being merely men, they endeavour to survey the universe from the standpoint of gods. There is a pretence at adequate clarity of fundamental ideas. We can never disengage our measure of clarity from a pragmatic sufficiency within occasions of ill-defined limitations. Clarity always means "clear enough."

With this preamble, we now turn to the papers read this afternoon. It is out of place to discuss them in detail at the close of this session. They will remain in my mind as a landmark for future thought. Also, where they indicate difficulties, I am in general agreement as to the need of clarification or revision in my written works. Of course you will not expect an adequate exposition of philosophy in thirty minutes.

John Dewey asks me to decide between the "genetic-functional" interpretation of first principles and the "mathematical-formal" interpretation. There is no one from whom one more dislikes to differ, than from Dewey. William James and John Dewey will stand out as having infused philosophy with new life, and with a new relevance to the modern world. But I must decline to make this decision. The beauty of philosophy is its many facets. Our present problem is the fusion of two interpretations. The historic process of the world, which requires the genetic-functional interpretation, also requires for its understanding some insight into those ultimate principles of existence which express the necessary connections within the flux.

For example, there are meaningful relations between these ten fingers, and the billions of stars, and the innumerable billions of atoms. The interrelations of the specific multiplicities of groups of individual things constitute the clearest example of metaphysical necessity issuing in meaningful relations amid the accidents of history. No explosion of any star can generate the multiplication-table by any genetic-functioning. But such functioning does exemplify interrelations of number. It is necessary that the meaning of the explosion be partly expressed by arithmetic. This necessity underlies the accidents of the explosion.

By a queer chance in this epoch of the universe arithmetical patterns constitute some of the clearest insights of human intelligence. There are limits to this clarity. But such as it is, we teach it to infants. Metaphysical knowledge enters while we still remember the rocking cradle. The notion of "many things" carries with it the necessity that there be numbers. And yet there is no necessity that any special relationship of numbers be in any one instance exemplified. In this way we can observe the curious interweaving of accident and necessity.

The notion of "many things" is a slippery one. There are these ten fingers and there are the ten commandments. In what sense do these fingers and the ten commandments together constitute twenty things? We are here brought up against the difficulty of the subtle change of meaning in familiar notions according to the context in which they occur.

The vagueness of our insight prevents our exact understanding of the metaphysical basis of particular exemplification. For this reason our metaphysical notions are an approximation. They represent such disengagement of necessity from accident as we are able to attain. One illustration of this approximate character of metaphysical knowledge is that such knowledge is always haunted by alternatives which we reject. Now necessity permits no alternatives. A century ago, arithmetic as then understood seemed to exclude alternatives. To-day, the enunciation of ultimate arithmetic principles is beset with perplexities, and is the favourite occupation of opposing groups of dogmatists. We have not yet arrived at the understanding of arithmetical principles which exhibits them as devoid of alternatives.

Plato's ultimate forms, which are for him the basis of all reality, can be construed as referring to the metaphysical necessity which underlies historic accident. In the case of his immediate successors, the superior lucidity of arithmetic insight triumphed. The result was that the Academy after his death tended to identify the forms with arithmetic notions. Indeed the Academy and subsequent European philosophers went further. They saw in Euclidean Geometry another example of necessity. We now know that they were wrong. The continual breakdown of pretensions to the achievement of final metaphysical truth is pathetic. But, on the other side, the persistent presupposition of final principles cannot be neglected by any philosopher who counts himself as a "radical empiricist." For example, to take John Dewey's language in his paper which is spread before me, the compound word "genetic-functional" means an ultimate metaphysical principle from which there is no escape. I am here in complete agreement with Dewey. The idea is vague, and adumbrates something beyond exact definition. This vagueness arises because Plato and Dewey are men with the limitations of human insight.

This notion of human limitations requires guarding. There is an implicit philosophic tradition that there are set limitations for human experience, to be discovered in a blue-print preserved in some Institute of Technology. In the long ancestry of humans, from oysters to apes, and from apes to modern man, we can discern no trace of such set limitation. Nor can I discern any reason, apart from dogmatic assumption, why any factor in the universe should not be manifest in some flash of human consciousness. If the experience be unusual, verbalization may be for us impossible. We are then deprived of our chief instrument of recall, comparison, and communication. Nevertheless, we have no ground to limit our capacity for experience by our existing technology of expression.

Thus to say that human experience is limited is not

to assert a standard limitation for all occasions of all humans. There are usual limitations depending on that dominant social order of our epoch, which we term the Laws of Nature and the habits of humanity.

This vagueness is not due to a morbid craving for metaphysics. It haunts our most familiar experiences. Consider the following set of notions:—The weight of that man: The height of that man: The intelligence of that man: The kindness of that man: The happiness of that man: The identity of that man with his previous self yesterday.

In the first place, the exact meaning of "that man"—body and soul—would puzzle the wisest to express. Yet each phrase is sufficiently clear for inexact common sense. Secondly, the small inconspicuous words in various phrases seem to alter their meaning from phrase to phrase. In the above examples, consider the little word "of." There is nothing about it alarmingly metaphysical. My small dictionary gives as its first meaning "Associated or connected with." I suggest to you that "weight," and "height," and "intelligence," and "kindness," and "happiness," and "self-identity with a previous existence," are each of them "associated or connected with" a man in its own peculiar way. Thus in each phrase the word "of" has changed its meaning from its use in the other phrases. Yet, after all, there is a fundamental identity underlying all these changes; and the pompous phrase "associated or connected with" is the best that the dictionary can do in the way of reminding us of that fact.

This conclusion has an important bearing on Logic. Consider the phrase "S is P." This proposition is a way of drawing your attention to "the P-ness of S," either for the sake of belief, or for some other purpose. If we neglect the irrelevant psychological accompaniments in the production of this phrase, we see that the word "is" in "S is P" reproduces the meaning of the word "of" in "the P-ness of S." Thus the meaning of "is" varies with changes in S or in P.

But an argument consists in a preliminary grouping of propositions, together with a deduction of other

propositions. Thus in addition to the criticism of the
original propositions as to truth or falsehood, we
require a criticism as to whether the undoubted changes
of meaning, in the same word appearing in different
propositions, are relevant to the argument. Also as new
propositions are deduced the same criticism is required.
Thus the simple-minded notion of logical premises vanishes. The little words "is," "and," "or," "together," are
traps of ambiguity.

Of course, gross common sense can usually settle the
matter. But experience has shown that as soon as you
leave the beaten track of vague clarity, and trust to exactness, you will meet difficulties. I remember when
Bertrand Russell discovered his well-known paradox.
He sent it by letter to Frege who was then alive. Frege's
answering letter began with the sentence, "Alas, arithmetic totters."

One source of vagueness is deficiency of language. We
can see the variations of meaning; although we cannot
verbalize them in any decisive, handy manner. Thus we
cannot weave into a train of thought what we can apprehend in flashes. We are left with the deceptive identity of the repeated word. Philosophy is largely the effort
to lift such insights into verbal expression. For this
reason, conventional English is the twin sister to barren
thought. Plato had recourse to myth.

The method of algebra embodies the greatest discovery for the partial remedy of defective language. The
procedure of the method is to select a few notions of
the simplest interconnections of things; such connections, for example, as are expressed by the words "is,"
"of," "and," "or," "plus," "minus," "more than," "less
than," "equivalent to," and so on indefinitely. A small
group of such terms is selected, on the principle that
expressions containing them are again capable of interconnection by these same notions.

The fundamental assumption is that these basic connectives retain an invariable meaning throughout the
algebraic development of patterns, and of patterns of
patterns. The legitimacy of this assumption is guarded

by the device of the "real variable," as it is termed. Symbols, such as the single letters, p, q, r, x, y, z, u, v, w, are used under the assumption that each symbol indicates one and the same individual thing in its repetitions throughout the complex pattern. Also it is assumed that the things represented yield meaningful patterns as thus connected. Also it is assumed that the inevitable variation of meaning infused into these basic symbols of interconnection by the diversity of the variable is not such as to affect that meaning which the pattern contains for the observers in question.

There are thus four fundamental assumptions, namely: (1) The invariableness of the basic terms of interconnection (the connectives), (2) the invariableness of the unspecified entities indicated by the symbols for "real variables," (3) the meaningfulness of the patterns of real variables, thus connected, (4) the irrelevance to the argument of the completion of meaning infused into the basic connectives by the unspecified real variables thus connected. Namely, the meaning as in assumption (1) is not in fact invariable, but the variation is irrelevant.

These principles of algebraic symbolism express the concurrence of mathematical formal principles with accidental factors. This concurrence is inevitable for the production of meaningful composition. And apart from composition there is no meaning, that is to say, there is nothing. The clarity is deceptive, as the clash of the first and fourth assumptions shows. Finally we are forced back to the pragmatic justification—It works, And yet it "totters," unless care be taken.

The basic connectives are the relevant mathematical-formal principles. The real variables are the unspecified accidental factors. But the connection of the accidents is not a mere mathematical-formal principle. It is the concrete accidental fact of those accidents as thus connected. This suffusion of the connective by the things connected is the most general expression of the genetic-functional character of the universe. It also explains the vagueness which shrouds our metaphysical insight. We are unable to complete the approximation of disengag-

ing the principles from the accidents of their exemplifications.

Necessity requires accident and accident requires
necessity. Thus the algebraic method is our best approach to the expression of necessity, by reason of its
reduction of accident to the ghost-like character of the
real variable.

It follows from this explanation of the algebraic
method, that our mathematics and our symbolic logic,
as hitherto developed, represent only a minute fragment
of its possibilities. In making this statement I shelter
myself behind a quotation (*Sophist* 253 CD):

> *Stranger.* Now since we have agreed that the classes
> or genera also commingle with one another, or do not
> commingle, in the same way, must not he possess some
> science and proceed by the processes of reason [he]
> who is to show correctly which of the classes harmo-
> nize with which, and which reject one another, and
> also if he is to show whether there are *some elements
> extending through all and holding them together so
> that they can mingle, and again, when they separate,
> whether there are other universal causes of separation.*
>
> *Theaetetus.* Certainly he needs science, and perhaps
> the greatest of sciences.

Also to Plato we can add the authority of Leibniz.
And now having invoked such support, I can cease the
defence of the attempt to bring together the genetic-
functional and the mathematical-formal methods in one
philosophic outlook.

Philosophic thought has to start from some limited
section of our experience—from epistemology, or from
natural science, or from theology, or from mathematics.
Also the investigation always retains the taint of its
starting point. Every starting point has its merits, and
its selection must depend upon the individual philosopher.

My own belief is that at present the most fruitful,
because the most neglected, starting point is that section of value-theory which we term æsthetics. Our enjoyment of the values of human art, or of natural beauty,

our horror at the obvious vulgarities and defacements which force themselves upon us—all these modes of experience are sufficiently abstracted to be relatively obvious. And yet evidently they disclose the very meaning of things.

Habits of thought and sociological habits survive because in some broad sense they promote æsthetic enjoyment. There is an ultimate satisfaction to be derived from them. Thus when the pragmatist asks whether "it works," he is asking whether it issues in æsthetic satisfaction. The judge of the Supreme Court is giving his decision on the basis of the æsthetic satisfaction of the harmonization of the American Constitution with the activities of modern America.

Now there are two sides to æsthetic experience. In the first place, it involves a subjective sense of individuality. It is *my* enjoyment. I may forget myself; but all the same the enjoyment is mine, the pleasure is mine, and the pain is mine. Æsthetic enjoyment demands an individualized universe.

In the second place, there is the æsthetic object which is identified in experience as the source of subjective feeling. In so far as such abstraction can be made, so that there is a definite object correlated to a definite subjective reaction, there is a singular exclusive unity in this æsthetic object. There is a peculiar unity in a good pattern.

Consider a good picture. It expresses a unity of mutual relevance. It resents the suggestion of addition. No extra patch of scarlet can be placed in it without wrecking its unity.

The point is that the subjective unity of feeling and the objective unity of mutual relevance express respectively a relation of exclusion to the world beyond. There is a completion which rejects alternatives. Mere omission is characteristic of confusion. Rejection belongs to intelligible pattern.

This doctrine extends, or distorts, the meaning of another saying of Plato, when he says that not-being is a form of being. Here I am saying that rejection is a form of prehension. But I fully agree with Dr. Ushenko that this doctrine requires examination, and probably should be recast. However, I adhere to the position that it is an approximation to an important truth.

We must end with my first love—Symbolic Logic. When in the distant future the subject has expanded, so as to examine patterns depending on connections other than those of space, number, and quantity—when this expansion has occurred, I suggest that Symbolic Logic, that is to say, the symbolic examination of pattern with the use of real variables, will become the foundation of æsthetics. From that stage it will proceed to conquer ethics and theology. The circle will then have made its full turn, and we shall be back to the logical attitude of the epoch of St. Thomas Aquinas. It was from St. Thomas that the seventeenth century revolted by the production of its mathematical method, which is the re-birth of logic.

The result of our human outlook is the interweaving of apparent order with apparent accident. The order appears as necessity suffused with accident, the accident appears as accident suffused with necessity. The necessity is, in a sense, static; but it is the static form of functional process. The process is what it is by reason of its form, and the form exists as the essence of process.

To hold necessity apart from accident, and to hold form apart from process, is an ideal of the understanding. The approximation to this ideal is the romantic history of the development of human intelligence.

My relation to Hegel's philosophy has, I hope, been made plain by this paper. He is a great thinker who claims respect. My criticism of his procedure is that when in his discussion he arrives at a contradiction, he construes it as a crisis in the universe. I am not so hopeful of our status in the nature of things. Hegel's philosophic attitude is that of a god. But I must leave Hegel to those who have studied him at first hand.

Uniformity and Contingency

THE SUBJECT MATTER which I propose to consider in this paper is a well-worn theme of philosophy, and I cannot hope in any essential way to remove the difficulties which encompass it. My endeavour will be merely to restate the problem with attention to distinctions and discriminations which are sometimes insufficiently emphasized.

The general problem is to examine, whether any isolated portion of our experience has any character which of itself implies a corresponding character, extending beyond the domain of that immediate example. In other words, we ask whether, on the ground of experience, we can deduce any systematic uniformity, extending throughout any types of entities, or throughout the relations between them. Where uniformity ends, contingency commences. The whole subject has been discussed by Hume in his *Philosophical Essays concerning Human Understanding,* with a clarity which constitutes his investigations a classic locus, from which all subsequent discussion must start. In order to get the discussion under way, I will start with some quotations from Hume:—

> "An annalist or historian, who should undertake to write the history of Europe during any century, would be influenced by the connexion of contiguity in time and place. All events, which happen in that portion of space, and period of time, are comprehended in his design, tho' in other respects different and unconnected. They have still a species of unity,

141

amidst all their diversity" (Essay III, of the *Association of Ideas*).

" 'Tis universally allowed by modern enquirers, that all the sensible qualities of objects, such as hard, soft, hot, cold, white, black, etc., are merely secondary and exist not in the objects themselves, but are perceptions of the mind, without any external archetype or model which they represent. If this be allowed, with regard to secondary qualities, it must also follow with regard to the supposed primary qualities of extension and solidity; nor can the latter be any more entitled to that denomination than the former. The idea of extension is entirely acquired from the senses of sight and feeling; and if all the qualities, perceived by the senses, be in the mind not in the object, the same conclusion must reach the idea of extension, which is wholly dependent on the sensible ideas or the ideas of secondary qualities" (Essay XII, of the *Academic or Sceptical Philosophy*).

I wonder whether this was one of the passages which awoke Kant from his dogmatic slumber. He certainly accepts the argument by his doctrine of space and time as forms of intuition.

Hume accepts, without question, space and time as reigning throughout nature. It is in fact the very basis of his celebrated analysis of the idea of necessary connexion amongst events. He says: "It appears, then, that this idea of necessary connexion amongst events arises from a number of similar instances, which occur, of the constant conjunction of these events, nor can that idea ever be suggested by any one of these instances, surveyed in all possible lights and positions. But there is nothing in a number of instances, different from every single instance, which is supposed to be exactly similar; except only, that after a repetition of similar instances, the mind is carried by habit, upon the appearance of one event, to expect its usual attendant and to believe, that it will exist.

"This connexion, therefore, which we *feel* in the mind, or customary transitive of the imagination from one object to its usual attendant, is the sentiment or impression, from which we form the idea of power or necessary connexion" (Essay VII, of the *Idea of Necessary Connexion*).

You will notice that in this passage "the constant conjunction" of events and the "attendance" of one event on another must mean spatio-temporal contiguity, or else the whole point of his explanation of the idea of causation is lost. Accordingly the spatio-temporal character of nature is a presupposition of Hume's philosophy. I am not making any objection to Hume's assumption: far from it, I am claiming his support. What Hume says of the history of Europe is true of any set of events. "They have still a species of unity, amidst all their diversity." They obtain this "species of unity" in virtue of their joint inclusion within some definite four-dimensional region of space and time.

I ask now, on what basis do we ground the assumption of the spatio-temporality of nature? The presupposition stands on a different basis to contingent occurrences. If time and space cease to be, there is a rupture in the texture of experience; but when the Campanile in Venice collapsed, the incident was unexpected and regrettable, but did not otherwise affect the intrinsic character of things observed. The status of space and time is in some way different from that of the Campanile.

In the absence of space-time there may still be consciousness aware of the truths of pure mathematics. It so happens that in fact we contemplate these mathematical truths in a temporal succession. But this order of precedence in our consideration of mathematics seems casual and irrelevant, so that we can easily imagine a timeless mathematical knowledge. In the same way the idea of a spaceless mathematical knowledge presents no difficulty; and mathematics, as thus known, would even contain the science of pure geometry, viewed as an abstract mathematical subject. Accordingly we cannot maintain that knowledge in itself requires space-time, either as conditioning the mode of consciousness, or as an essential system of relations interconnecting the things known.

Again we cannot maintain that the mode of apprehension, in consciousness, of a spatio-temporal nature requires that the mode itself should include temporal transition. In other words, the fact, that nature is a process, does not require that consciousness of nature should be a process. For the moment of consciousness involves a specious present in which there are antecedents and consequents. Accordingly, such process as in fact does attach to consciousness is not the necessary consequence of the apprehension of process. For if this were the case, the suspension (relatively to the process of nature) of the process of consciousness, so as to include the specious present in immediate apprehension, would be impossible; consciousness would have to put the past behind it, in step with nature.

Accordingly, by an indefinite enlargement of the specious present, we can imagine an awareness of all nature as a process, although no process is implicated in the mode of awareness. Accordingly we can dismiss the process of consciousness as irrelevant to the immediate enquiry, and can concentrate on the fact, that nature, as apprehended in consciousness, is constituted as a process, and that the analysis of this constitution is expressed by the properties of space-time.

The peculiarity of the space-time process is, that any part of it establishes the whole scheme within which the remainder is set. We can imagine that, in the realm of existence, there may be an alternative space-time process other than that of nature; but nature and the alternative process do not conjoin to make one process. In fact we are aware of such alternative processes in dreams, where we apprehend a process of events which in respect to nature are nowhere and at no time. The dating of the dream is the correlation of the process of the apprehending consciousness with the space-time of nature. But, in respect to the matter of the dream, fortunately there is no region of nature which was the field of those awful events. Let it be noted that the new relativity doctrine has a vital connexion with the theory of dreams. According to the older views it was open for an objector to say that the dream-date of the dream-events was the real time of night as correlated with the process of consciousness, but that the dream-space and its dream-contents were imaginary. But space and time have now been assimilated, so that you cannot tear them apart. Accordingly, when the dream-space is assigned to an imaginary world, so is the dream-time. It therefore becomes necessary to distinguish between the process of apprehension and the apprehended process.

The distinction between the dream-world and nature is, that the space-time of the dream-world cannot conjoin with the scheme of the space-time of nature, as constituted by any part of nature. The dream-world is nowhere at no time, though it has a dream-time and a dream-space of its own. We may ask anyone who, in contradiction to this doctrine, maintains the contingency of space-time relations, untempered by any uniformity imposed by any single part or region, how the dreams are to be discriminated from natural occurrences. The course of nature is entirely contingent, since Hume's doctrine merely explains the growth of our expectation and has no reference to the actual course of nature in the future. Suppose that one morning you wake up in your bedroom, having dreamt that you were tossed by a bull. You know that it was a dream, because here you are, safe in bed, and you dined and went to bed quietly last night. Also you recollect that, when you went to bed, you had not been tossed by a bull. Accordingly it must have happened during the night. Why should it not have happened just as really as your dinner or breakfast? There is no saying what will be the course of events, and your experience now shows you, that you may be tossed by a bull during the night, and be none the worse next morning.

Why not? Hume speaks of vivid impressions and faint copies. But a good nightmare is as vivid an impression as most of us ever have. Also you may say that in a dream we do not notice subsidiary circumstances. But this omission surely cannot discriminate dreams from reality. Anyone tossed by a bull, either in or out of a

dream, is in a bad position to take notes of the landscape. Many people in ordinary conditions fail to note
the most obvious circumstances. It is said that a Prime
Minister's wife was once in terror lest her husband
should have gone to the House of Commons without
his trousers, so unobservant was he of subsidiary details.
Further, in dreams we often take very careful notes of all
details. I remember once having the dream of hovering,
and in my dream taking the most careful notes. I remembered that I had had the experience before, and I
had subsequently decided that it was a dream. Accordingly, I decided to observe all the circumstances with
great exactness, so as not again to be led into disbelief
by my vague recollection of the details. When I woke
I remembered a vivid experience, with all its details
carefully observed. Unfortunately, it would not fit into
the space-time framework of my waking experience. But
otherwise there was nothing against it.

Thus the position we are led to is that we are aware
of a dominant space-time continuum and that reality
consists of the sense-objects projected into that continuum. It is not true that the apprehended process invariably fits into this dominant continuum: for example,
dreams do not. But it is true that by an indirect inference we can always correlate the process of apprehension with the dominant continuum: for example, in the
case of a dream we can note the time of going to bed
and the time of waking, and can correlate the process
of apprehending the dream with some portion of the
intervening night. An apprehended process which does
not fit in with the dominant continuum is called imaginary, and its status must be considered separately. It
is probably true that a vague sense of the dominant
process even persists through sleep, but it cannot be re-
called as a distinct recollection.

The fitting in of distinct apprehended processes into
one dominant continuum—for example, my life in the
morning with my life in the afternoon of the same day
—can only mean that the apprehended process of the
morning has disclosed a scheme of relations amid relata,

which extends beyond itself (*i.e.,* beyond my life of the morning), so that my experience of the afternoon is nothing else than the apprehension of a process which is included in this predetermined scheme, and it is apprehended as being thus included. The same explanation holds of the continuity of the apprehended process of my life for shorter periods, from hour to hour, from minute to minute, and from second to second. If the spatio-temporal continuity does not mean this, what does it mean? Furthermore, if there be no apprehended spatio-temporal continuity of this character, how do the advocates of experience as our sole source of knowledge propose to exclude dreams from the realm of reality? You may put it in this way: A standard of normality, independent of arbitrary selection, is essential in the philosophy of experience.

It is not necessary to maintain, and it probably is not true, that awareness of a dominant space-time continuity is necessary for consciousness. It seems very improbable that such consciousness as appertains to the lowest type of conscious animals includes any such awareness. It seems more likely that a delicate sense for spatio-temporal continuity, with its accompanying discrimination of reality from illusion, is the last product of a developed consciousness. It is certainly easily destroyed or weakened, and its loss is compatible with rationality and some measure of sense apprehension.

Accordingly, our awareness of nature consists of the projection of sense-objects—such as colours, shades, sounds, smells, touches, bodily feelings—into a spatio-temporal continuum either within or without our bodies. In fact our bodies are primarily the loci within this continuum of the special class of sense-objects which I have called bodily feelings. But "projection" implies a sensorium which is the origin of projection. This sensorium is within our bodies, and each sense-object can only be described as located in any region of space-time —say, in any "event"—by reference to a particular simultaneous location of a bodily sensorium. We cannot say that a colour is in such-and-such a position at such-and-such a time without referring to some definite sensorium with some simultaneous location, for which it is true. Accordingly, the process of projection consists in our awareness of an irreducible many-termed relation between the sense-object in question, the bodily sensorium, and the space-time continuum, and it also requires our awareness of that continuum as stratified into layers of simultaneity, whose temporal thickness depends on the specious present.

I have suggested the term "ingression" for this many-termed relation. Accordingly, I would say, that we are aware of the ingression of sense-objects amid the events of a dominant space-time continuum, and that this awareness constitutes our apprehension of nature.

If this account of nature be accepted, then space-time must be uniform. For any part of it settles the scheme of relations for the whole, irrespective of the particular mode in which any other part of it, in the future or the past or elsewhere in space, may exhibit the ingression of sense-objects. Accordingly, the scheme of relations must be exhibited with a systematic uniformity. Thus (to repeat), the discrimination of reality from dream requires an apprehended dominant space-time continuum, determined in its totality, and this determination requires that it be uniform. We have here the primary ground of uniformity in nature.

There is another line of thought by which this same conclusion can be reached. I have developed it in the James-Scott lecture, delivered before the Royal Society of Edinburgh, and included in my book, *The Principle of Relativity*.[1] This argument is based on broader, and to that extent firmer, ground than the discussion here given. It proceeds from the consideration of the status of any particular item of knowledge, variously called a "factor of fact" or an "entity." It claims that their embeddedness in an all-embracing fact is essential for their very being, so that in this sense all particulars are abstractions. Fact is not another entity, but is the general all-embracingness of reality. There is then an argument,

[1] Cambridge University Press, 1922.

not here reproduced, that correlative to the significance of each factor for fact, there is the patience of fact for each factor, and that this patience must exhibit itself as a systematic uniformity within fact. The argument given above is a particular application to the more general argument of that lecture.

Before proceeding to develop consequences from this conclusion that nature is the observed field of this relationship of ingression, I must consider two objections which may be produced to my preceding argument. Hume, it may be said, provides a standard of normality by reference to what is usual, so that, according to him, the repeated impact of the usual on our minds automatically produces a judgment according to this standard. In fact, the essence of Hume's doctrine is our expectation of the usual. I have already quoted his own statement of this doctrine, and will now repeat it: "But there is nothing in a number of instances, different from every single instance, which is supposed to be exactly similar; except only that, after a repetition of similar instances, the mind is carried by habit, upon the appearance of one event, to expect its usual attendant, and to believe, that it will exist."

I am myself accepting Hume's doctrine, and am merely investigating the presuppositions which it involves. My point is, that this doctrine will not suffice to discriminate dreams from actual occurrences. Some dreams are very usual, and some occurrences are very rare. For example, my dream of hovering has been much more usual in my experience than my first-hand experiences of glaciers. Why (on Hume's principle) should I turn my hoverings out of nature, and retain my excursions on glaciers? Surely it is very arbitrary. But it may be said that other people have been on glaciers, and it is their concurrent testimony which we trust. I am afraid that, if you read Hume carefully, this argument will not hold. I do not understand how other people's experience can "carry" my mind "by habit." Furthermore, it is probable that among the twelve hundred million people now existing, not to speak of previous ages, there have occurred many more dreams of hovering than excursions on glaciers. Indeed, I do not know how to conduct such an extensive census of other people's experience, and still less do I see how to obtain the information in time to make it of use in the quick bustle of daily life.

Furthermore, I suspect that the tabulated results would be very disconcerting. I am inclined to believe that the majority of humankind do include dreams among the events of nature. There is a tomb somewhere—in Cairo, I think—of a Mahometan saint. The ascription of the tomb to the saint is peculiarly certain, because an angel took someone there in a dream, and showed him the spot. Does not that belief represent the attitude of the majority? Your only ground for scepticism (assuming the good faith of the dreamer) must be that, by direct inspection of your own dreams, you see that their space-time is incoherent with your dominant space-time, and that therefore you suspect the same of other people's dreams.

I pass now to the second objection. It is urged that sense-objects—to us the term which I have applied to colours, sounds, bodily feelings, and such like things— are purely individual and mental, and that the common nature, in which we are incarnate, and which is the nature described in science, is a different order of being from these psychological offshoots of mental excitement. I again draw your attention to Hume, who has stated to perfection the first comment to be made on this doctrine. I repeat the passage which I have already quoted: " 'Tis universally allowed by modern enquirers, that all the sensible qualities of objects, such as hard, soft, hot, cold, white, black, etc., are merely secondary, and exist not in the objects themselves, but are perceptions of the mind, without any external archetype or model, which they represent. If this be allowed, with regard to secondary qualities, it must also follow with regard to the supposed primary qualities of extension and solidity; nor can the latter be any more entitled to that denomination than the former. The idea of extension is entirely acquired from the senses of sight and feeling; and if all the qualities, perceived by the senses, be in the mind, not in the object, the same conclusion must reach the idea of extension, which is wholly dependent on the sensible ideas or the ideas of secondary qualities."

But, according to the new relativity theory, space and time cannot be disjoined. Thus—if we follow the line of thought of the objection—not only must perceived space, but also perceived time, be considered as mental and purely personal to each individual. But we have agreed that all our knowledge is based on experience. We are thus led to the conclusion that all our knowledge is the play of our own mind. Indeed, on this supposition, it is a mere silly trick which leads me to speak in the plural, and I cannot imagine how I acquired the habit. For I have no source of information to give me news of anything beyond myself. The space-time of science is thus absolutely swept away.[1]

My own position is that consciousness is a factor within fact and involves its knowledge. Thus apprehended nature is involved in our consciousness. But in its exhibition of this character our consciousness exhibits its significance of factors of fact beyond itself.

I differ from the idealists, so far as they consider such an external significance as peculiar to consciousness and thence deduce that the things signified have a peculiar dependence on consciousness. I ascribe an analogous external significance to every factor of fact, such as the colour green or a bath-chair. Correlative to the significance of nature by consciousness, there is the patience of consciousness by nature. Nature exhibits the fact, that it is apprehensible by consciousness. The ingression of sense-objects amid events is a character of nature exhibiting this patience. Also the stratification into layers of simultaneity, which is an essential character of this ingression, is at the same time an adaptation of nature, so that our finite consciousness of it is possible, and is also an adaptation of consciousness for the apprehension

[1] I have considered this line of thought more in detail in my *Concept of Nature*, Camb. Univ. Press, 1920, under the heading "The Bifurcation of Nature."

of nature. In other words, it is both a fact of nature, and is also the way in which we apprehend nature. In separately abstracting consciousness and nature from their embeddedness in all embracing fact, each exhibits its patience of the other.

The space-time continuum is not the sole basis of uniformity in nature. If it were so, induction would be impossible. It is here that we find the weakness in Hume's, and in some other, philosophies. Hume explains a ground for the origin of our instinctive trust in induction. But unfortunately his explanation does not disclose any rational explanation of this trust. The rational conclusion from Hume's philosophy has been drawn by those among the lilies of the field, who take no thought for the morrow. Hume admits this conclusion. He writes:—

"The sceptic, therefore, had better keep in his proper sphere, and display those philosophical objections, which arise from more profound researches. Here he seems to have ample matter of triumph; while he justly insists, that all our evidence for any matter of fact, which lies beyond the testimony of sense or memory, is derived entirely from the relation of cause and effect; that we have no other idea of this relation than that of two objects, which have been frequently conjoin'd together; that we have no arguments to convince us, that objects, which have, in our experience, been frequently conjoin'd, will likewise, in other instances, be conjoined in the same manner; and that nothing leads us to this inference but custom or a certain instinct of our nature; which it is indeed difficult to resist, but which, like other instincts, may be fallacious or deceitful" (Essay XII, of the *Academic or Sceptical Philosophy*).

Hume runs away from his own conclusion: he adds:—

"On the contrary, he (a Pyrrhonian) must acknowledge, if he will acknowledge anything, that all human life must perish, were his principles universally and steadily to prevail" (*loc. cit.*).

I wonder how Hume knows this: it must be that there is some element in our knowledge of nature which his philosophy has failed to take account of. Bertrand Russell adopts Hume's position. He says:—

"If, however, we know of a very large number of cases in which A is followed by B and few or none in which the sequence fails, we shall in *practice* be justified in saying 'A causes B,' provided we do not attach to the notion of cause any of the metaphysical superstitions that have gathered about the word" (*Analysis of Mind,* Lecture V, Causal Laws).

Again I should like to know how Russell has acquired the piece of information which he has emphasized by italics—"we shall in *practice* be justified, etc."

I do not like this habit among philosophers, of having recourse to secret stores of information, which are not allowed for in their system of philosophy. They are the ghost of Berkeley's "God," and are about as communicative.

I do not conceive myself to have solved the difficulty which puzzled Hume. But I wish to point out the direction in which, as I believe, the complete solution will be found. In an extract, already quoted, he has stated the issue with his usual clearness:—

"But there is nothing in a number of instances, different from every single instance, which is supposed to be exactly similar; except only, that after a repetition of similar instances, the mind is carried by habit, upon the appearance of one event, to expect its usual attendance, and to believe, that it will exist."

Hume's philosophy found nothing in any single instance to justify the mind's expectation. Accordingly he was reduced to explaining the origin of the mind's expectation otherwise than by its rational justification. It follows, that, if we are to get out of Hume's difficulty, we must find something in each single instance, which would justify the belief. The key to the mystery is not to be found in the accumulation of instances, but in the intrinsic character of each instance. When we have found that, we will have struck at the heart of Hume's argument.

This overlooked character of the single instance must be its significance of something other than itself. This extra something will thus be known by relatedness, arising from the knowledge of the single instance by adjective. We have already found, that the spatio-temporal significance of each single instance is a necessary presupposition of Hume's whole philosophy of nature. We have now to ask, whether there is not some further significance.

There obviously is this further significance. For the single instance is an instance of the ingression of sense-objects amid events. But every sentient being passes at once to the perceptual objects indicated by that instance. How do we pass from the ingression of sense-objects to the perceptual objects? The answer is that the ingression signifies the objects. It is no good saying, that the accumulation of instances of "smell and a pat" reminds a dog of his master by the association of ideas. Hume's argument applies: If no one instance is significant of his master, but is merely a smell and a pat, what virtue towards producing the master can the accumulation possess? The significance may grow clearer to perception by the accumulation of instances, but it must have been there from the beginning.

A perceptual object is a true Aristotelian adjective of some event which is its situation. It is what I have elsewhere (cf. Principle of Relativity, Chapters II and IV) called a "pervasive" adjective—meaning by that term an adjective of an event which is also an adjective of any temporal slice of that event. For example, a perceptual object—say, a chair—which has lasted in a room for one hour, has also lasted in the room during any one minute of that hour, and so on. A sense-object has also in general the pervasive property; but its relation to its situation is entirely different from that of a perceptual object, in that it is derived from its ingression in nature, which is an irreducible many-termed relation.

The point, which I am maintaining, is that the ingression of a sense-object into nature is significant of perceptual objects, so that thereby perceptual objects are known by relatedness. I have previously argued that this ingression is significant of the space-time continuum. But, of course, I do not mean that there are detached independent significances. The ingression is significant of events which are characterized by pervasive Aristotelian adjectives. The event is not bare space-time which is a further abstraction. An event is qualified space-time—or rather, the qualities and the space-time are both further abstractions from the more concrete event.

This significance of ingression is, in respect to space-time, more vividly exhibited by the reference of the sense-object to its situation. But, in respect to perceptual objects, it is more vividly exhibited by the reference, inherent in the ingression, to a sensorium (or percipient event) "here," which is recognized in consciousness as its seat in nature. This sensorium is an event—roughly, the body or part of the body—qualified by an Aristotelian pervasive adjective. Furthermore, where the sense-object is a bodily feeling, there is a peculiar vividness of recognition of parts of the body as perceptual objects, in that the vivid reference to the sensorium is now used with the fainter, vaguer reference, of the sense-object to a perceptual object in its situation.

But, where the sense-object has its situation projected beyond the body, a difficulty arises. Undoubtedly there is reference to a perceptual object. You see a candle, where the candle is a perceptual object. But this reference to a perceptual object—other than the sensorium— is apt to be vague, illusive or absent. You see double; you see the image behind the looking-glass; you hear stray sounds vaguely filling the space around you; you smell a scent.

The reference of the sense-object to the perceptual object is not as neat as we should desire for simplicity of exposition.

The sense of touch gives a peculiarly vivid reference,

and for that reason has been taken as the standard of verification. Doubting Thomas wished to touch his Lord. A vivid reference is also obtained by an accumulation of sense-objects of different types, whose various ingressions relegate them to the same situation.

The evidence is summed up in the statement that the ingression of sense-objects into nature involves events analysable into space-time qualified by pervasive Aristotelian adjectives. The sensoria are always indicated in this way as the loci of perceptual objects, and also in general so are the situations of the sense-objects. But what are the perceptual objects—tables, trees, stones, etc.—which are thus signified? For unbiased evidence of their character we must have recourse to the general popular idea, and not to scientific accounts, elaborated in the interest of theories, and vitiated by faulty analyses of nature. The popular evidence is unanimous:— The modes of ingression of sense-objects in nature are the outcome of the perceptual objects exhibiting themselves. The grass exhibits itself as green, the bell exhibits itself as tolling, the sugar as tasting, the stone as touchable.

Thus the ultimate character of perceptual objects is that they are Aristotelian pervasive adjectives which are the controls of ingression.

Now an Aristotelian adjective marks a breakdown of the reign of relativity; it is just an adjective of the event which it qualifies. And this relation of adjective to subject requires no reference to anything else. Accordingly, a perceptual object is neutral as regards events, other than those which it qualifies. It is thus sharply distinguished from a sense-object, whose ingression involves all sorts of events in all sorts of ways.

Furthermore, the contingency of ingression, with its baffling tangle, is now simplified into the contingency attaching to the simpler relations of perceptual objects to the events which they qualify.

But, if the very nature of perceptual objects is to be controls, have we not in them those missing characters of events, whose supposed absence led Hume to remove causation from nature into the mind? A control is necessarily the control of the process, or transition, in finite events. It thus means, in its essential character, a control of the future from the basis of the present. Thus in modern scientific phraseology, a perceptual object means a present focus and a field of force streaming out into the future. This field of force represents the type of control of the future exercised by the perceptual object —which is, in fact, the perceptual object in its relation to the future, while the present focus is the perceptual object in its relation to the present. But the present has also a duration. What we observe is the control in action during the specious present.

There are a finite number of perceptual objects within any region of space-time relevant to our experience. This finiteness still remains as we pass from the somewhat vague perceptual objects to the more precise scientific objects such as electrons. Accordingly, there are a finite number of such controls of the future, which are in any way relevant to our experience.

The latest and subtlest analysis of the difficulties which cluster round the notion of Induction is to be found in Part III of J. M. Keynes's *Treatise on Probability*. I will conclude with a quotation from his profound discussion:—

> "The purpose of the discussion, which occupies the greater part of this chapter, is to maintain that, if the premises of our argument permit us to assume that the facts or proposition, with which the argument is concerned, belong to a *finite* system, then probable knowledge can be validly obtained by means of an inductive argument." (*Treatise on Probability*, Ch. XXII.)

PART III

EDUCATION

The Study of the Past—Its Uses and Its Dangers

FOR EACH SUCCEEDING generation, the problem of Education is new. What at the beginning was enterprise, after the lapse of five and twenty years has become repetition. All the proportions belonging to a complex scheme of influences upon our students have shifted in their effectiveness. In the lecture halls of a university, as indeed in every sphere of life, the best homage which we can pay to our predecessors to whom we owe the greatness of our inheritance is to emulate their courage.

In allusion to the title of this address—The Study of the Past, Its Uses and Its Dangers—I may at once say that the main danger is the lack of discrimination between the details which are now irrelevant and the main principles which urge forward human existence, ever renewing their vitality by incarnation in novel detail.

THE PRESENT A TURNING POINT IN WESTERN CIVILIZATION

It so happens that the first five and twenty years of the existence of the Harvard Graduate School of Business Administration exactly coincides with a turning point in the sociological conceptions of Western Civilization. Here, by the term Western Civilization, I mean the sociological habits of the European races from the

Ural Mountains on the boundary of Asia passing westward half way round the world to the shores of the Pacific Ocean, that is from 60° east longitude to rather more than 120° west longitude.

If you keep to the northern temperate zone, in every country that you can pass through in this long journey you will find some profound agitation, examining and remodelling the ways of social life handed down from the preceding four hundred years. This agitation as a major feature in social life is the product of the past twenty-five years. Of course this unrest has its long antecedents, but within this final short period the disturbance has become dominant. Undoubtedly, something has come to an end.

It is also worth noticing that the centre of disturbance seems to lie within each country. We are not dealing with the repercussion of a revolution with one local centre. In Russia there has been a revolution, because something has come to an end. In Asia Minor the Turks are recreating novel forms of social life, because something has come to an end. Throughout Central Europe, every nation is in a ferment, because something has come to an end. With one exception in the larger nations of Western Europe, Italy, Spain, Germany, England, there is a turmoil of reconstruction, because something has come to an end. The one exception is France. In that country, the internal motive seems to be absent—perhaps fortunately for her. But anyhow, the comparative absence of any feeling of the end of ways of procedure explains a certain inability to penetrate instinctively into the springs of action of her neighbouring nations. For the rest of Europe, something has come to an end; while France is prepared to resume the practice of traditions derived from Richelieu, from Turgot, from the French Revolution, and above all from her incomparable craftsmen.

When in this survey we cross the Atlantic and come to America, I do not think that there is exaggeration in the refrain, that something has come to an end. We stand at the commencement of a new thrust in sociological functioning, and this novelty is of supreme importance in respect to the education of our future leaders in business administration. Do not misunderstand me. In each nation we all want to continue to aim at our old ideals. We can only preserve the essence of the past by the embodiment of it in novelty of detail. I will anticipate the argument by stating my belief that the best feature in the past was the sturdy individualism fostered by the conditions of those times. I am here referring to the last two centuries in the life of America, of England, and of Continental Europe. Why I have drawn attention to the universality of the present sickness is to draw the conclusion that the remedy is not to be found in the adjustment of some detail peculiar to any one nation. In each nation there will be details of change peculiar to it, and between nations there will be differences of proportion. But we are not likely to recognize the necessary group of details unless we have some grasp of the general character of the disease.

THE PRECEDING TREND FROM MEDIÆVALISM TO INDIVIDUALISM

What has come to an end is a mode of sociological functioning which from the beginning of the sixteenth century onwards has been slowly rising to dominance within the European races. I mean that trend to free, unfettered, individual activity in craftsmanship, in agriculture, and in all mercantile transactions. The culmination of this epoch, with its trend still in this direction, can be roughly assigned to the stretch of time from the middle of the eighteenth century to the middle of the nineteenth century. During that hundred years the populations in Europe and America suffered many evils from want, starvation and war. These evils have always afflicted mankind in the mass. Within this period one essential quality stimulated all sociological functionings.

That quality was hope—not the hope of ignorance. The peculiar character of this central period was that the wise men hoped, and that as yet no circumstance had arisen to throw doubt upon the grounds of such

hope. The chief seats of economic, and of general sociological, speculation were in France and Great Britain.
The realized trend towards individualism, and away
from mediævalism, had vastly simplified the problem
of constructing a social theory which should correspond
to the practical ideal of civilized life when relieved from
the madness of its traditional rulers, Kings, Priests, and
Nobles. For nearly three hundred years before the middle of the eighteenth century, a continual process of
simplification in practice and in theory had prevailed.
Feudalism was in full decay, the complex interweaving
of church authority with secular government was steadily vanishing. Society could be conceived as functioning
in terms of the friendly competition of its individual
members, with the State standing as umpire in the minority of instances when there occurred a breakdown of
these normal relations.

Primarily this competition of activities concerned the
production of material goods for the support of physical
life. As to other values, the later formula of "A Free
Church in a Free State" sufficed. There the word
"Church" suggests religion. But it was in practice extended to all organizations for the supply of every variety
of non-material values, religious, æsthetic, moral, including the natural feelings of human affection. I am endeavouring to sketch to you the perfect doctrine of an
individualistic society, which was naïvely presupposed
in sociological theorizing from the midst of the eighteenth century to that of the nineteenth. The doctrines
were never realized in their full purity. But all social
progress was in the direction which they indicated. These
doctrines were more perfectly realized in America than
elsewhere. But they also admirably fitted themselves to
the needs of the commercial middle classes in England,
France, and wherever in Europe this middle class was
a chief factor in the social life. The American Revolution and the French Revolution were dramatic incidents
arising from the acceptance of this sociology. The reconstruction of Europe after the Napoleonic Wars was
guided by it. Also, it was evidently the fact that life

was healthier, finer, more upstanding, in proportion to the dominance of this social individualism.

In respect to this doctrine, Where do we stand to-day? I will not quote from any theorist indulging in brilliant speculation. I will take a sentence from the editorial page of one of the leading Boston newspapers, in its current issue which has been placed by my side as I write. Here is the judgment of this organ of Boston commercial opinion—"Whether we fancy it or not, we are in the midst of a revolution, so far as concerns the relation of the individual to the Federal Government."

Evidently something has happened. The pure milk of the word of the sociological Gospel, perfected in the late eighteenth century, has gone sour.

Undoubtedly, during the central period the Gospel of Individualism was working well wherever it was tried. But only in North America was it ever the wholly dominant fact. In Europe, it always had the aspect of a new mode of sociological functioning gradually superseding the relics of an antecedent order. This older layer of law and custom had somewhat the air of a deposit of rubbish, in process of removal. Perhaps this aspect was partly a mistake. The relics of the older order may have been providing for the realization of a diversity of values which the pure practice of the new sociology would have left unsatisfied. I will return to this point later.

In America very special conditions for human life were at that time in full operation. An empty continent, peculiarly well suited for European races, was in process of occupation. Also that section of these races which felt the urge towards that type of human adventure had freely selected itself to constitute the American population devoted to this enterprise. Accordingly, in America this epoch exhibits a wonderful development of sturdy independence, with the individual members of the population freely carving out their own destinies. This is the Epic Epoch of American life, and after the initial struggles of small beginnings it had a wonderful central period of about a hundred and fifty to two hundred years. It was a triumph of individual freedom, for those who liked that sort of opportunity. And the population was largely selected by its own or its ancestral urge towards exactly that sort of life. Indeed the evil side of the survivals of feudalism in European life is illustrated by the bitter feelings which lingered amid the recollections and traditions of the American population. This episode in human existence, when individualism dominated American life, cannot be too closely studied by sociologists. It is the only instance where large masses of civilized mankind have enjoyed a regime of unqualified individualism, unfettered by law or custom.

THE NEW FORCES AT WORK—THE PASSING
OF INDIVIDUALISM

From the middle of the nineteenth century, new forces have been at work, and gradually the situation both in Europe and America has been changed. Up to that time, for nearly two centuries human progress had been identified with the advance of individualism. In England, the Industrial Revolution had been in operation for about seventy years, and in America and Continental Europe for a somewhat shorter period. Its first effect had been to promote the sturdy individualism of the middle classes. It enriched them and stimulated their energy. It destroyed the decaying elements of the past. About a hundred years ago two Englishmen were leaning on the balustrade of a railway bridge, watching a railway train pass under them. This was a novel sight in those days. "It is an ugly thing," said one of them, "but it is the death of Feudalism." The speaker was a strong advocate of the liberal individualism characteristic of that epoch. He did not foresee that in another two generations the new mechanism would send the then existing Individualism into the same grave with the old Feudalism.

Of this trilogy, Feudalism, Individualism, Ugliness, to-day the Ugliness alone survives, a living threat to the values of life.

The recent phase of modern industrialism has been

produced by a change of scale in industrial operations. One of the dangerous fallacies in the construction of scientific theory is to make observations upon one scale of magnitude and to translate their results into laws valid for another scale. Almost always some large modification is required, and an entire inversion of fundamental conceptions may be necessary. For example, on a large scale of observation there are bits of matter, such as rocks, tables, lumps of iron, solid, resistant, immobile. On another, microscopic, scale there are a welter of molecules in ceaseless activity and each molecule only definable in terms of such activity. The physical science of the two preceding centuries made exactly this mistake. It naïvely transferred principles derived from its large-scale observations to apply to the operations of nature within the minute scale of individual atoms. I suggest that our sociological doctrines have made the same error in the opposite direction as to scales. We argue from small-scale relations between humans, say two men and a boy on a desert island, to the theory of the relations of the great commercial organizations either with the general public or internally with their own personnel. In any one corporation we may have to consider tens of thousands of employees, hundreds of executives, scores of directors, scores of thousands of owners, and a few controlling financial magnates in the background. I am not saying that such corporations are undesirable. That is not my belief. Indeed, such organizations are necessary for our modern type of civilization. But I do say that observations of the behaviours of two men and a boy on a desert island, or of the inhabitants of a small country village, have very little to do with the sociology of our modern type of industrial civilization.

In any large city, almost everyone is an employee, employing his working hours in exact ways predetermined by others. Even his manners may be prescribed. So far as sheer individual freedom is concerned, there was more diffused freedom in the City of London in the year 1633, when Charles the First was King, than there is to-day in any industrial city of the world.

It is impossible to understand the social history of our ancestors unless we remember the surging freedom which then existed within the cities, of England, of Flanders, of the Rhine Valley, and of Northern Italy. Under our present industrial system, this type of freedom is being lost. This loss means the fading from human life of values infinitely precious to it. The divergent urges of different individual temperaments can no longer find their various satisfactions in serious activities. There only remain iron-bound conditions of employment and trivial amusements for leisure. I suggest that one subject of study for our industrial and sociological statesmen should be the preservation of freedom for those who are engaged in mass production and mass distribution which are necessities in our modern civilization. It is a study requiring penetrating insight so as to distinguish between the realities of freedom and its mere show, and between hurtful and fruitful ways of freedom.

My point is that the change of scale in modern industry has made nearly the whole of previous literature on the topic irrelevant, and indeed mischievous. The study requires deep consideration of the various values for human life. I am not suggesting any facile solution. The topic is very perplexing. It involves many branches of psychology—general psychology, industrial psychology, and mass psychology. It involves sociological and political theory. It involves the rôle of æsthetics in human contentment. It involves an estimate of the sense of effectiveness aroused by co-operation in enterprises with large aims. It involves the understanding of physiological requirements. It involves presuppositions, however dim, as to the aims of human life. But above all, and beyond all, it involves direct observation and practical experience. Unless the twentieth century can produce a whole body of reasoned literature elucidating the many aspects of this great topic, it will go hard with the civilization that we love.

THE INCREASING PRESSURE OF GENERAL UNEMPLOYMENT

I now turn to another consideration which cannot be separated from the previous topic. The dangers to freedom are largely cloaked in times of prosperity by the scarcity of labour. In such times, at least the desire for change can be satisfied. If one job does not suit, a man can try another. If the type of work remains the same, at least there is a change of factory, or overseers, and of associates. The real cause for restlessness may lie deeper, but something has happened. Also the scarcity of labour affects the mentality of the management.

In a time of widespread unemployment this outlet for discontent is closed. A man is lucky if he be not in the bread line. There is a very real closing down of freedom for everyone concerned, from the higher executives to the lowest grade of employees. In any industrial district in the world to-day, it is a grim joke to speak of freedom. All that remains is the phantasm of freedom, devoid of opportunity.

It is therefore of the first importance to have in our minds some estimate of the probable frequency of these periods when there is an excess of labour. In England, where long ago any pioneering period has ceased, this excess of labour is almost normal. Sometimes it is more so, and sometimes less. But for more than a hundred years—indeed from the time of Queen Elizabeth—the out-of-work problem has been always there. There are reasons why this evil should settle on the whole industrial world as a permanent factor in life, unless the great corporations can adapt their mode of functioning. Up to now the problem has been mitigated by the existence of empty continents in the temperate zones. This relief has vanished.

The combination of mass production and of technological improvement secures that more and more standardized goods can be produced by fewer and fewer workers. Here and there there are mitigating causes.

But the general fact remains, ever advancing in importance. The issue is unemployment. The proper phrase is "technological unemployment." But you do not get rid of a grim fact by the use of a technical term. The result is that a portion of the population can supply the standardized necessaries of life, and the first luxuries, for the whole population. A portion of the population will be idle, and as time goes on, this portion will grow larger.

In the second place, the demand for goods grows slack. This is to be expected. For the idle, or the partially idle, cannot afford to be brisk buyers. Thus the full quota of goods for the whole population is not wanted, and so again there is another reason for unemployment.

These two grounds for dullness of trade are often cloaked by other agencies with a contrary effect. They stay, however, permanently in the background, a constant aggravation of any trade depression, and a constant provocation of bitter discontent.

THE MISTAKEN POLICY OF MODERN SALESMANSHIP AND PRODUCTION

Beyond these effects, the modern salesmanship associated with mass production is producing a more deep-seated reason for the insecurity of trade. We are witnessing a determined attempt to canalize the æsthetic enjoyments of the population. A certain broad canalization is, of course, necessary. Apart from large uniformities, all effort is ineffective. But all intensity of enjoyment, sustained with the strength of individual character, arises from the individual taste diversifying the stream of uniformity. Destroy individuality, and you are left with a vacancy of æsthetic feeling, drifting this way and that, with vague satisfactions and vague discontents.

This destruction is produced by the determined attempt to force completely finished standardized products upon the buyers. The whole motive appealed to is conformation to a standard fashion and not individual satisfaction with the individual thing. The result must be the creation of a public with feeble individual tastes.

There is nothing that they really want to buy, unless the world around them is also buying. This is an admirable condition for mass buying when the times are favourable. But it is an equally effective condition for mass abstention from buying when expenditure is once checked. The stimulus of the individual want for the individual thing has been destroyed. And after all, individual buyers have to buy individual things.

Thus large oscillations in demand arise from comparatively slight causes. The sturdy individuality of the mass of the public is the greatest security for steadiness in trade. It is this individuality that the great commercial corporations are setting themselves to destroy. Independent individuality of demand is the tacit presupposition of much of the older political economy. Thus, whether we survey the producers or the buyers, we find the same steady decay of individuality. Now, a decay of individuality finally means the gradual vanishing of æsthetic preferences as effective factors in social behaviours. The æsthetic capacities of the producers and the æsthetic cravings of the buyers are losing any real effectiveness. The canalization of the whole range of industry is in rapid progress. Apart from the dangers of economic prosperity, there is in this decay a loss to happiness. Varied feelings are fading out. We are left with generalized mass emotion.

My line of argument up to this point is not the preliminary to an attack on great commercial corporations. These organizations are the first stage of a new and beneficent social structure. My complaint is that in the two or three generations of their existence on their present scale, they have functioned much too simply. They should enlarge the scope of their activities. To understand what is required, we should ask why France stands out as a tremendous exception to the general sociological trend. She has preserved the individuality of her craftsmanship and the individuality of her æsthetic appreciations. This is the secret of the undying vitality of the French nation. What I suggest is that the great corporations in various ways should interweave in their organizations individual craftsmanship operating upon the products of their mass production. For example, take the most obvious of all the æsthetic products in the world to-day, namely, the dress of women. If you enter a leading Boston store to-day, you are lucky if you find material illustrating as many as five shades of blue. The other shades are out of fashion. Such is the decree of the business world. Delicate craftsmanship, subtle combinations, individualities of taste are out of the question. To-day the world of women is restricted to the fashionable blue. Subtlety of taste is ruthlessly stamped out. In this example, the delicacies of craftsmanship are irrelevant to the operations of the great producing firms.

Of course, mass production underlies the modern standards of life. What we require is a close interweaving of the two forms of activity, the production of the general material and the perfection of the individual thing. Of course there are all sorts of half-way houses. I can only here state my meaning in crude outline. The great producers and the great distributing corporations should include in their activities the work of craftsmen and designers.

THE NEED FOR ECONOMIC STATESMANSHIP

This concerns another point relating to our present troubles. We are told that the seat of the evil is our distribution of goods. In its more obvious sense this doctrine is certainly wrong. The organization for the delivery of goods to any purchasers could hardly be more perfect, whether we think of the neighbourhood of a town, or of the whole of America, or of the whole of the civilized world. What is at fault is that for the majority of people the ability to procure goods depends upon the exercise of some useful activity. But mass production has restricted the quantity of human activity required. Hence a large number of people are unable to procure goods. But when we look at France, we see that we have wantonly suppressed a very considerable side of human activity, for which large numbers of human beings are admirably fitted. The result is that we have both suppressed the opportunity for self-expression which is so necessary for happiness, and have extinguished the claim of a large section of people upon the goods which lie ready for consumption.

What is defective is not distribution, but the variety of opportunity for useful activity. Thus the interweaving of mass production with craftsmanship should be the supreme object of economic statesmanship. Here by craftsmanship I do not mean the exact reproduction of types of activity belonging to the past. I mean the evolution of such types of individual design and of individual procedure as are proper for the crude material which lies ready for fashioning into particular products.

Also I must avoid another misconception. Nothing that I am saying has any reference to any action now desirable for the rescue of the world from its present state of miserable depression. Such action must be immediate and must therefore presuppose the existing modes of economic activity. I am speaking of the tasks awaiting economic statesmanship during the next twenty-five years. My point is that in our economic system as now developed there is a starvation of human impulses, a denial of opportunity, a limitation of beneficial activity—in short, a lack of freedom. I have endeavoured to show that this fault in our system produces in various ways an excess of irritability in the social organism. Whatever system we have, the natural fluctuations of the universe will produce in it ups and downs, from better to worse. But this irritability latent in our present modes of functioning seizes hold of these fluctuations and exaggerates them. These recurrent depressions have been growing more and more dangerous and are likely to grow worse. In our search for remedies, we must consider the things that cannot be reduced to machinery, things material and things spiritual. Again, the present argument is only a putting together of considerations which have already been developed by others. In fact, much of this discussion is a commonplace of literature.

More than a hundred years ago, Southey pointed out the destruction of beauty in the lives of the operatives in the manufacturing districts of England. In the third quarter of the nineteenth century Ruskin was denouncing the absence of æsthetic values in the English industrial system. Both Southey and Ruskin were fantastic and impracticable, but they fastened attention upon a real blot. Relevantly to the present situation, Professor Ames of Dartmouth has expounded with great insight the case of interweaving æsthetic activities in the leisure provided by mechanized mass production. Also at Harvard, Professor Mayo and his band of workers have made a notable advance in the analysis of industrial psychology. In the combination of these points of views, we have the foundation for a new chapter in Economic Statesmanship.

THE EDUCATIONAL PROBLEM

This conclusion has a moral for education. A training in handicraft of all types should form a large element in every curriculum. Education is not merely an appeal to the abstract intelligence. Purposeful activity, intellectual activity, and the immediate sense of worth-while achievement, should be conjoined in a unity of experience. Of course, this doctrine must be worked with discretion, and in proportion to the other necessities of education. At the latest stage of education, namely in university life, a differentiation takes place. A large proportion of the students should devote themselves to sheer intellectual training. But in Science, in Technology, and in Art, a large infusion of hand work should be a serious element in the work of a considerable section of the students. My own experience, which is a large one, in the educational requirements of the population in London has convinced me that the sharp distinction between institutions devoted to abstract knowledge and those devoted to application and to handicraft is a mistake. Every university will have its emphasis, this way or that. But I see no advantage in an anxious drawing of an exact line of demarcation. The mass of mankind, including many of its most valuable leaders, requires something betwixt and between. Common sense and no abstract theory should dictate what any particular university attempts.

This doctrine is most easily exemplified in the sphere of the fine arts, though, of course, craftsmanship is not limited to artistic production. Industry and art alike will be on a healthier basis when the natural avenue to fine art is through craftsmanship with the cultivation of fine sensibility. In addition to the general design of the mass-made product, there can also be interwoven possibilities of adding individual differentiation to individual things. The finer genius can develop into the specialized artist with his work abstracted from all association with other utility.

In conclusion, it is obvious that any blending of a machine age with a vigorous craftsmanship will require a large co-operation between schools and universities and the great business interests concerned with production and distribution. It will also require the education of the general public. It will require the advice of technologists of all types from engineers to artists. It will destroy much of the sweet simplicity of modern business policy which fastens its attention solely on one aspect of our complex human nature. But if it can be accomplished it will add to the happiness of mankind, notably so by stabilizing the popular requirements and widening the area of useful occupations.

Education and Self-Education

THERE HAVE BEEN moments in history when new worlds were discovered. There was such a moment when Columbus discovered America. Creation widened to man's view. There is such a moment now. We are all aware that the immediate future holds within it possibilities different from anything that has been known in the past. Our views are widened.

Mankind has entered upon a new phase. It is no good saying that you will go on in the future as you have done in the past. You can't; and it is the reason for this impossibility which I want to talk about this afternoon, and to connect with the life of this institution, whose admirable work we are here to celebrate.

One great reason is the war, perhaps the greatest tragedy which mankind has ever passed through in so short a time. This is an obvious cause for immediate unsettlement, which no one can forget. But there have been wars and revolutions before; and, yet, often enough, human life has fallen back into the same old groove of habit. The great war has been the occasion for the suddenness of the change, but it only disclosed slow-working forces whose combined action was just reaching a new importance. For month after month you may shake a tree, and nothing much will happen. At length the fruit is ripe, and one storm will bring it down. The social systems of the European races were ripe for change, when war swept across them. The effective agents in this preparation for change are, the growth of science, the growth

174

of invention, the growth of industrial organization, the extension of education, and the demand that opportunity for rational existence should be shared by the whole community. Evidently, these agencies depend on each other, and it is the resonance due to their joint action which makes the tidal wave which is sweeping us along. I see no cause for alarm, and much for rejoicing, provided that we understand and shape our actions with courage.

The rapid advance of scientific theory is a recent factor in human history, a matter of the last few centuries. Previously, the way to set about physical investigations had not been clearly realized. Also, so little was known that it was difficult to find out more; because the more you know, the easier it is to add to your knowledge. Thus, centuries ago mankind had to wait for the rare inspirations of genius. To-day we can organize first-rate talent for research, and scientific theory steadily grows at an increasingly rapid rate. A little genius doesn't hurt; but we are not stationary between the flashes.

This knowledge of nature has opened a boundless field for inventions of practical utility in industry. Again, we have only recently found out that there is a method in invention, and that the foundation of this method is a thorough training in science. A hundred years ago no one could have anticipated how important this ceaseless search after new methods, new processes, and new machines would be. But we need not go back a hundred years. Only five years ago the average British attitude was to trust to luck, and to hope that no disturbing invention would come our way in the hands of a rival.

The last five years have taught us many things. Behind the armies in the field there has raged a ceaseless battle of inventors. Aeroplanes, antidotes to poison gases, methods of detecting submarines, explosives, engines, magnetos, optical glass, tanks, wireless telegraphy—these names only represent some of the matter in which invention has been essential for victory or at least for a while staved off ultimate defeat.

A well-known chemist told me that in the early days

of poison gases, he tried twenty-seven different antidotes before he hit on the one that succeeded. And such investigations had to be repeated again and again as the gases in use varied. Unless the Germans had known how to use the nitrogen of the air in the formation of nitrates for high explosives and for fertilizers, our blockade would have finished them off in the first three years. If we had ended the war with the same types of aeroplanes and with the same equipment for them as we had at the beginning, we should have been defeated, and this afternoon you would have been obeying the orders of Prussian officers.

The war has afforded the most astounding proof of what can be effected by an intense concentration upon invention, provided that you utilize the trained ability. Now the same effect can be produced in most departments of industry, not merely for a short period, but continuously—granted that your science advances, and you train the men to apply it. Now you cannot think that these lessons have been overlooked by other nations. Of course, they haven't. The Americans, the French, the Italians, the Germans, the Japanese, will concentrate inventive faculty upon every detailed process of their manufactures. In the State of the future, to use obsolete industrial methods will be little short of a criminal offence. However, it will be unnecessary for the police to prosecute, for bankruptcy will automatically overtake the offenders.

Thus the old conception of industry in a static state, forever employing the old methods and producing the old goods, will not fit the facts. Managers, designers, and artisans must be equipped to adapt themselves to circumstances which are ever changing. Now, to produce this adaptability there is only one method, and that is education. This education must not be conceived on narrow lines. It is of no use to train the young in one very special process which will probably be superseded before they are middle-aged. Give them alert minds exercised in observation and in reasoning, with some knowledge of the world around them, and with feeling for beauty. Then a sound training in handicraft will be accompanied with a power of adaptation and a natural love of efficiency. This is the way to produce a happy people of high capacity for production. It does not require any great gift of prophecy to foresee that that nation will have in effect won the war, which most clearly learns this lesson of the war. If you let your education fall behindhand, no heroism can save you.

And here I must remind you of the fine story of the Eastern king, who, in a vision, chose wisdom, because for its own sake he preferred it to all the treasures of Oriental magnificence. Power follows wisdom, because nature unlocks its secrets to the wise and dowers the temperate with zest and energy. Wisdom should be more than intellectual acuteness. It includes reverence and sympathy, and a recognition of those limitations which bound all human endeavour.

A nation won't get wisdom except by the love of it. And it is here that the modern democratic demand for a due share in the opportunities of life is full of hope and of anxiety. Of course, the demand is of mixed origin, for it is human. It gains its moral energy from the ultimate rights of the moral and intellectual natures of man, his right to his own creative actions directed by his own wisdom—a right based on an insatiable craving for what gives worth to existence.

This cry for freedom seems at times to sleep for ages, like the fire in a volcano. When it wakes, the day of God's judgment has arrived, and the worth of human societies is being weighed in His scales. Those societies perish which exhibit mainly selfishness and cowardice. Courage and hope are your best armour with which to meet a revolution—and, above all, mutual sympathy.

You have endured the hardships and dangers of war, fighting side by side. Death showed no favour, and in the hour of danger you were brothers fighting for the freedom of the world. Let the love that bound you together in the terrible struggle you have just come through guide and counsel you now in your struggle for the birth of a just state of society.

We have good reason for hope, based upon the most critical survey of our prospects. The true effect of an invention is always that a given amount of manual labour and of nervous energy, and of brain-capacity, will produce a greater amount of the material goods which mankind requires. Thus the progress of invention should mean the greater productivity of labour. Furthermore, it is the high grade labour, which will increase its productivity in the greatest ratio. It is only this stamp of labour which, in the long run and on a large scale, can be trusted to do its job quickly, accurately, and punctually. I am sure that the facts are with me, if you examine any department of industrial life.

But high-grade labour requires high-grade management. You cannot make a greater mistake than to think that as the status of labour rises, the need for management declines. The exact contrary is the case. The higher the status of the workman, the more precious he becomes, and the less can any waste of his time or ineffective use of his work be tolerated. Now it is an encouraging fact that recent years have seen a decided improvement in the general management of our industrial system. During the war the internal management of factories was revolutionized, with the result that though wages went up, the cost per shell went down. Furthermore, the external organization of industry is steadily improving, namely, the arrangements for buying the raw material, and for marketing the finished products. The ultimate reason is the improvement in means of communication— telegraphy, ships, and railways in the past, and, in the future, also aeroplanes. I am told that the immediate use of aeroplanes commercially will be for the quick sending of samples.

The world really is ready for a new step forward. The demand for better conditions is not a crying for the moon; but its satisfaction depends upon a bold trust in the value of education. Mr. Fisher spoke the exact truth in the House of Commons when he said that it is not a question whether we can afford so much education, but whether we can afford so little. Ignorance is the most expensive of all luxuries, with its wild oscillations from apathy to impatience—an impatience which ignores the iron limitations set by nature.

Now, what is the meaning of all this education about which we hear so much? A large number of my audience are in the midst of their own education, and may be supposed, therefore, to know something about it. Things do not always look quite the same when you are in the midst of them, and when you are outside them. In the midst of a cloud it appears as a dull grey fog; at some distance it may be iridescent with the brilliance of the setting sun. I rather wonder whether education does not appear to me clothed more brightly than it does to some of you.

First, as to its meaning, the word "education" means literally, the process of leading out. Thus we are talking of the way in which all your faculties and capacities should be encouraged to expand and unfold themselves. Consider how nature generally sets to work to educate the living organisms which teem on this earth. You cannot begin to understand nature's method unless you grasp the fact that the essential spring of all growth is within you. All that you can get from without is some food, material or spiritual, with which to build your own organism, and some stimulus to spur you to activity. What is really essential in your development you must do for yourselves. The regular method of nature is a happy process of genial encouragement. There can be very little satisfactory growth with the exclusion of this method.

Think of the world on a brilliant spring morning. All the educational reforms of recent years come from realizing the profound truths conveyed in this scene of joyous growth. Any educational authority which forgets this is throwing half its money into the dusthole. Neither mind nor body can develop satisfactorily in a strait-waistcoat. Those children who told their new nurse that they were the sort who wanted treating with kindness, had grasped the true philosophy of education. The first thing that a teacher has to do when he enters the class-room is to make his class glad to be there. Now, so far, I expect that my theory of education will receive the enthusiastic approval of the younger members of my audience.

This doctrine of enjoyment bears decisively on the meaning of a liberal education. Such an education is not characterized by its subjects, or by the number of subjects. It all depends on how they are treated. Whatever creates a disinterested curiosity for knowledge, or an appreciation of beauty, enlarges the mind and causes it to expand by its own free inward impulse. The natural mode of intellectual stimulus is by action and reaction between our immediate actions and our immediate thoughts. You remember the great text of scripture, "How shall a man love God whom he has not seen, if he love not his brother whom he has seen?" This text explains the comparative failure of pretentious systems for liberal education.

At this point I wish emphatically to commend the careful co-ordination of the studies in this school. I mean the way in which the different tasks set to the boys play into each other's hands. His mathematics enables him to understand the science lessons; and the mathematics and the science together enable him to understand what he is doing in the workshop where he is engaged in the metal work and the woodwork. The pleasure in handicraft comes when what you are doing is expressing your own ideas. We all know that when we talk. A conversation when we express our own ideas is in general less tiring than reading aloud a printed page which expresses some other person's thoughts. In education you double the rate of progress of your pupils if you can fit their various studies together; mankind is born for action; it is the very breath of his life that he should be doing something. The aim of education is the marriage of thought and action—that actions should be controlled by thought and that thoughts should issue in action. And beyond both their is the sense for what is worthy in thought and worthy in action.

In the democracy of the future every man and every woman will be trained for a free intellectual life by an education which is directly related to their immediate lives as citizens and as workers, and thereby elicits speculations and curiosities and hopes which range through the whole universe.

In such an education everything depends on the teachers and on the pupils. We are discovering that in schools you cannot do without genius, genius of character, genius of insight, and genius of intellectual enthusiasm. Authorities who want successful schools must see to it that the conditions in the teaching profession are those in which genius can thrive.

But, at the end of it all, you who are the pupils must bring your own enjoyment to your tasks.

Remember what I said a minute ago, that in reality you educate yourselves. No one else can do it for you. You are not pieces of clay which clever teachers are modelling into educated men. It is your own effort which alone essentially counts. So, finally, you have got to provide your own enjoyment by interesting yourselves in things which are worth doing and worth thinking about. Your working lives will either be a drudgery or a pleasure according to the way you take it. Of course, nothing can prevent your having long, grey times of hard work, tiring and disappointing; you can't take through anything without grit. Yet there is a great satisfaction in doing things skilfully, and in understanding all about what you are doing, and in thinking of how it all bears on the lives of others around you.

You won't get interested in what you are doing unless you have some ideals before you—some hopes for the betterment of human society, some joy in making others happy, some courage in facing the obstacles to progress. Such ideals bear essentially upon your school work. Ideals which are not backed by exact knowledge are mere fluffy emotion, and often lead to disastrous action. It is no use having vague ideals as to the importance of electrical machines. Each particular machine requires a definite design depending upon an exact knowledge of the purpose it is to serve, it requires a definite knowledge of the tools and material with which to make it; and before it can be used, a definite business organization must bring it to the notice of those who may find it serviceable.

Now, in all this there is nothing special to electricity or to machinery. England is a great democracy. You are going to manage your social clubs and your trade societies, and to be voters for local councils, and for the governing of your Empire. Surely you want some knowledge to tell you what men and women have thought, what they have felt, what they have enjoyed, and what they have suffered. You lay the foundation of this reading at school; you learn to read books there, and get to know the sort of things that there are to be known. Your real education comes in after life. I am not thinking of dull books. The best story books are some of your best teachers. A good novel or a good book of travel will let you know more of the world than many a treatise: only, for heaven's sake, think as you read. Try to imagine what it all means. Do not get a mere craving for print without thought. It is almost as bad as drink.

Then, there is the other side of your work, the science and the handicraft. The transformations of modern society depend upon these. Steam engines, dynamos, internal combustion engines, and machine tools in factories are, in their combined effect, more important agents in shaping lives of men than all the political theories since the world began. In short, do not be content with vague aspirations. Always push on to definite knowledge.

And you who are entering on life, and who in a few short years will form the England which has been saved by the heroism of our sailors and our soldiers, you should feel yourselves consecrated beings. The war has been fought to determine the future of the world; it has been fought that you, who are the first of the coming generations, may have your own lives in your own hands.

It is for you that sailors kept ceaseless watch amid the fogs of the North Sea; it is for you that men lined the trenches in Flanders amid dangers and discomforts indescribable; it is for you that the airmen dared their flights. For you the death, and for you the suffering. It is the promise of your lives which made it all worth while. And they succeeded. They kept the seas open; they drove the enemy from his strongest positions; and they made him surrender. But at what a cost in human suffering.

The memorial to these heroes must be built by you. It will last for thousands and thousands of years, and the life of each one of you is a stone in that building! The memorial is the future of the world, and that future will be determined by how your lives are fashioned.

Mathematics and Liberal Education

THE SUBJECT OF my address to-day is the consideration of the part which the elements of mathematics should play in a liberal education for the generality of boys up to the age of nineteen. The boys I mean are, of course, those who are capable of a liberal education. Wealth can do much, but it cannot give brains, and it cannot give character; nor can it give the intellectual interests which come from the union of brains with character. Accordingly, I exclude the residuum of boys, and am thinking of those only with fair brains and decent interests. Happily for England these constitute the great majority of the ordinary students who pass on to our Universities.

It will help us to understand the nature of our problem if we spend a few minutes in examining the ultimate reason for the existing upheaval in the scholastic world. We are, in fact, in the midst of an educational revolution caused by the dying away of the classical impulse which has dominated European thought since the time of the Rennaissance. I find it a little difficult to explain my meaning exactly. I am not referring to the mere teaching of a little more or a little less of Latin and Greek. What I mean is the loss of that sustained reference to classical literature for the sake of finding in it the expression of our best thoughts on all subjects. The Greek masterpieces of literature remain masterpieces of literature, and the labours of a band of brilliant scholars have reinterpreted them to the modern

184

world. But for all that, the scene presented to our view by the human life of to-day is essentially different to that presented either to the Greeks of two thousand years ago, or even to our grandfathers at the beginning of the nineteenth century.

There are three fundamental changes which make an unbridgeable gap. Science now enters into the very texture of our thoughts; its methods and results colour the imaginations of our poets; they modify the conclusions of philosophers and theologians. Again, mechanical inventions, which are the product of science, by altering the material possibilities of life, have transformed our industrial system, and thus have changed the very structure of Society. Finally, the idea of the World now means to us the whole round world of human affairs, from the revolutions of China to those of Peru. Even to our fathers it merely conveyed the idea of the nations of Europe, and, in particular, of the Mediterranean shores. But this provincial phase of thought is rapidly becoming impossible.

The total result of these changes is that the supreme merit of immediate relevance to the full compass of modern life has been lost to classical literature. To make a trivial example: in Athens, a reference to a potter's wheel might recall vivid memories of habitual sights; in London, it requires a footnote. Whether we regret it or no, the absolute dominance of classical ideas in education is necessarily doomed.

But it is not possible to alter the whole basis of our curriculum by a mere change in the time-table. In the first place, there is the difficulty which amateur reformers in education usually forget—that there is no time. Lack of time is the rock upon which the fairest educational schemes are wrecked. It has wrecked that scheme which our fathers constructed to meet the growing demand for the introduction of modern ideas. They simply increased the number of subjects taught. Latin and Greek were to be retained, the time given to mathematics enlarged, modern history and physical science to be added; and, at the same time, geography, music, drawing, and nature studies were not to be neglected. Also, of course, modern languages were regarded as indispensable.

The task of education with such a scheme of studies is frankly impossible. In all modern educational reform the watchword must be "concentration."

But if we examine this older curriculum as it existed in practice, we shall note, unless I am much mistaken, that in general it possessed one very striking peculiarity. It was believed, with some reason, that every cultivated idea had found its best expression in the classical literatures. The result was that the whole of the general training in ideas was annexed, and with some reason, to the study of the classical languages. Other studies were, in fact, pursued as mere technical acquirements. The boys might learn both German and Greek; but it was from Sophocles and not from Goethe that they drew their ideas. Mathematics, for example, was divested of all discussion of ideas, and reduced to the aimless acquirement of formal methods of procedure. In other words, modern thought was not introduced into the educational curriculum, but merely modern technique. If, for the mass of boys and young men, we are to concentrate our education upon modern subjects, we must first transform them into a real vehicle for the inculcation of ideas. We have, in fact, to civilize them.

Now, nothing is more difficult than to transmit to our pupils real general ideas, as distinct from pretentious phrases. Nobody with any sense can confront a class for long without discovering that all sound teaching is concerned with definite, accurate achievements on the part of the pupils—to construe grammatically a Latin sentence, to solve a quadratic equation, or to find, with some precision, the specific heat of lead. Vague generalities are worse than useless, and if we attempt to embody abstractions in short, precise formulæ, the pupils will simply learn them by heart as empty sounds.

In view of this difficulty, let us examine briefly how the classical languages achieved their undoubted success as vehicles of a liberal education. The advantage of education based upon them is that at every step definite aims are placed before the learner. He has to construe the author, to know the meaning and grammatical status of each word, and to render the sense in the precise equivalent English. There is nothing vague in all this; it is an accurate achievement which the pupil has to accomplish. It has also the useful property, which every teacher will appreciate, that it is easy to test whether the pupil has in reality tried to accomplish his task. He may not make sense of his translation, but he can at least know the meaning of the various words and their cases or their tenses; and, in addition to all this, the classical languages possess the supreme merit that great ideas are simultaneously presented to the mind. The noblest authors of Greece and Rome can be read. Some of us may still remember construing in our school-days Lucretius' reflections on the nature of the universe, and the account of the battle of the harbour of Syracuse, its triumph and its despair.

I will not consider further the question of the literary side of education. Not because I undervalue its importance, but because we are especially responsible for the logical training. We have really two general aims before us. In the first place, we have to teach what logic is. I do not mean by this that we should indulge in the somewhat futile task of affixing names to elementary logical processes after the manner of primers in formal logic. But we have to make our pupils feel by an acquired instinct what it means to be logical, and to know a precise idea when they see it; or, rather what unfortunately is more often wanted, to know an unprecise idea when they see it. In the second place, we have to make them understand that logic applies to life. This is, in fact, the harder task. Most people agree that there are abstract precise ideas capable of logical treatment, but very few really believe that a sensible man need take any account of them. Such and such ideas, they will say, are all right in theory, but in practice they are useless. It is here that the astounding success of modern science in transforming the world makes an examination of the elements of its logical methods so vital a part of modern education. In this region ancient thought is frankly useless. It is possible that the Greeks were in all respects abler than we are, and that, if here, they would conduct our scientific investigations in a manner superior to anything to which we can attain. But the ancient Greeks are not here, and the fact remains that our modern scientific thought completely overshadows anything of the same sort which existed in the ancient world. In this connection it is a mistake to think that the Greeks discovered the elements of mathematics, and that we have added the advanced parts of the subject. The opposite is more nearly the case; they were interested in the higher parts of the subject and never discovered its elements. The practical elements, as they are now employed in physical science, and the theoretical elements upon which the whole reposes, were alike unknown to them. Weierstrass' theory of limits and Georg Cantor's theory of sets of points are much more allied to Greek modes of thought than are our modern arithmetic, our modern theory of positive and negative numbers, our modern graphical representation of the functional relation, or our modern idea of the algebraic variable. Elementary mathematics is one of the most characteristic creations of modern thought. It is characteristic of modern thought by virtue of the intimate way in which it correlates theory and practice.

I make this point in no idle spirit, but to enforce a very serious conclusion. In the past, the teaching of elementary mathematics has suffered, not only because its ideas were sucked away from it by the dominant classics, but also because it was treated as a collection of mere uninteresting prolegomena to more advanced parts of the subject. But the mass of pupils never advanced to these further parts, and, in consequence, gained nothing but a set of purposeless dodges.

We must conceive elementary mathematics as a subject complete in itself, to be studied for its own sake. It must be purged of every element which can only be justified by reference to a more prolonged course of study. There can be nothing more destructive of true education than to spend long hours in the acquirement of ideas and methods which lead nowhere. It is fatal to all intellectual vitality. It produces, on the one hand, a sense of incompetence, of lack of grasp, and of inability really to penetrate to the true meaning of things; and, on the other hand, by a natural revolt of the self-respecting intellect, it produces a distaste for ideas, and a suspicion that they are all equally futile. I have had great experience with the average product of our schools as sent up to the Universities. My general conclusion is not that they have been idle at school, or have been taught carelessly. On the contrary, their education has evidently been supervised with a conscientious vigour. But there is a widely-spread sense of boredom with the very idea of learning. I attribute this to the fact that they have been taught too many things merely in the air, things which have no coherence with any train of thought such as would naturally occur to anyone, however intellectual, who has his being in this modern world. The whole apparatus of learning appears to them as nonsense. Of course, any individual schoolmaster is helpless in this matter; he is in the grip of the examination system. It is here that the utility of such associations as the one which celebrates its meeting to-day is apparent. It enables the results of first-hand experience to acquire the authority of a collective demand capable of constraining the nameless Furies who draw up our schedules of examinations.

But to return to elementary mathematics: we conceive it as a group of abstract ideas, and our course is to have a threefold character, namely: (1) the pupil is finally to be left with a precise perception of the nature of the abstractions acquired by constant use of them, illumined by explanations and finally by precise statements; (2) the logical treatment of such ideas is to be exemplified by trains of reasoning which employ them and interconnect them; and (3) the application of these ideas to the course of nature conceived in its widest sense as including human society is to be made familiar.

The subject, as thus broadly sketched out, is limited by the following considerations: there is very little time; only such ideas are to be introduced as are of fundamental importance to all mathematical reasoning, and they are not to be too complicated for the average boy to understand.

Most of these requisites explain themselves, and I need say nothing more about them. It will be seen that the whole spirit of these suggestions is towards a cutting down of the mere quantity of abstract reasoning to be performed, but towards an extension of the time devoted to a consideration of the ideas in themselves especially by the aid of their applications to examples. By examples, I mean important examples. What we want is one hour of the Caliph Omar, to burn up and utterly destroy all the silly mathematical problems which cumber our text-books. I protest against the presentation of mathematics as a silly subject with silly applications.

For example, take the theory of graphs, which, on its theoretical side, should teach the boys the abstract idea of a functional relation between variable quantities. This abstract idea is embodied in a few simple theoretical examples, such as the rectilinear graph of the linear algebraic function of one variable, the parabolic graph of the quadratic function, also the wavy graphs of the sine and cosine illustrate the general nature of periodic functions. In this way the boy grows familiar with the idea of an abstract precise law. If time permits the law of the inverse square can be exhibited in a graph, and also the fundamental law of the geometrical increase by plotting functions such as 2^x, 3^x, etc., and for the abler students by the consideration of the series $\exp\chi$, that is, e^x. But for the mass of boys a few well-chosen examples of precise functional relations would surely be better than a more ambitious course over a wider field.

Now, still keeping for the present to the abstract side of the work, the consideration of the zeros of these functions at once introduces equations as a necessary branch of study. Linear and quadratic equations acquire an important meaning, and so do the zeros of the sine and cosine functions. At this point I suggest the study of abstract algebra might well be stopped. I would utterly sweep away all prolonged multiplications and divisions, and the theories of greatest common measure and least common multiple, and complicated forms of linear and quadratic equations. They lead to nothing important in the boys' minds and consume a vast amount of valuable time. It would be quite sufficient to confine practice in multiplication to cases in which one factor is linear, and practice in division to cases in which the divisor is linear. Similarly, factorization, if admitted, should be rigidly confined to the case of two linear factors with the view of exemplifying the theory of the zeros of quadratic functions. Again, for the mass of boys, the algebraic treatment of fractional expressions consumes valuable time uselessly. Of course, ample practice in algebraic manipulation is necessary, but it should be restricted to a few types of the most necessary operations.

It will be noticed that while advocating the omission of a large part of the algebra usually taught, I would include the definitions of sine and cosine, and the study of their graphs. I do not suggest that trigonometry, properly so called, be introduced. By this I mean the application of the trigonometrical functions to the theory of the triangle. This would, in general, take up far too much time with very little intellectual result. The true use of these functions in elementary mathematics is their representation of the idea of periodicity. Perhaps, however, I am too sanguine in hoping that their comprehension is within the range of the ordinary boy.

The general outcome of these suggestions is that elementary algebra would be restricted to the consideration of the simplest functions of one variable. The golden rule should be that until the end of their course, students should never see expressions with more than one unknown in them. It is traditional at the very commencement of the study of the science to present functions containing a large number of letters, a, b, c, d, e, f, with directions to substitute particular values for them. I am utterly unable to conceive what is the educational value of this inane procedure. All that is wanted to begin with is the calculation of the values of particular cases of the simple functions for particular values of the single variable. Then the idea of any value of the function arises, and, to put the matter technically, we come to the letter "y" standing by itself on the left-hand side of the equation, and the function on the right-hand side.

Finally, we *perhaps* reach the idea of algebraic form and introduce the coefficients as parameters with the letters a, b, c. But except for the sake of algebraic form more than one variable, as an argument to a function, is never wanted in elementary study.

So much for the theoretical side of the subject. I have treated this first because in view of further suggestions on the practical applications, it was necessary to explain in what way time was to be gained, and what are the theoretical ideas to be led up to and to be applied.

But I wish emphatically to guard myself against suggesting that such abstract ideas as variable and functionality are best introduced by the consideration of abstract algebraic functions however simple. On the contrary, a preparatory consideration of concrete examples by graphical methods is surely necessary. There we reach one of the chief causes of the weakness of the traditional mathematical training. It is entirely out of relation to the real exhibition of the mathematical spirit in modern thought, with the result that it remained satisfied with examples which were both silly and unsystematic. Now the effect which we want to produce on our pupils is to generate a capacity to apply ideas to the concrete universe. Thus the examples which we choose form the very backbone of our teaching. The study of algebra should commence with a systematic study of the practical application of mathematical ideas of quantity to some important subject. Now what subject can we choose by which to represent the flux of quantities without the necessary intervention of algebraic technique from the very beginning. Many suggestions might be made, and it is obvious that many subjects in competent hands might be equally good. My suggestion in its crudest, and most aggressive, form is that half of the teaching of modern history should be handed over to the mathematicians. The phrase "handed over" is not quite accurate; for the half which I mean is the half which, although the true foundation of all knowledge of nations, is hardly taught. Our classical colleagues, excellent fellows as they are, have their limitations; and among them is this one, that they are not very fitted by their mental equipment to appreciate quantitative estimates of the forces which are moulding modern society. But without such estimates modern history as it unfolds itself before us is a meaningless tangle.

Now among other peculiarities of the nineteenth century is this one, that by initiating the systematic collection of statistics it has made the quantitative study of social forces possible. There are to our hands statistics of trade both external and internal, statistics of railway traffic, statistics of harvest, of prices, of health of population, of education, of crime, of income-tax returns, of national expenditure, of weather, of prices, of pauperism, and of times of sunrise and sunset throughout the year. The reduction of these to graphs, the careful study of the peculiarities of these graphs, the search for correlations among them, and the study of the public events which corresponded in time to peculiarities in graphical form, would teach more mathematics and more knowledge of modern social forces than all our present methods put together. Our relations with our colonies, with France, with Germany, with the United States, could be traced statistically. Problems could be set for solution by the boys—such as to state verbally the effect of war on the social life of a nation as exhibited by the graphs. Also they can be given the statistics and told to exhibit them in graphical form, and to state the general characteristics of the graphs thus obtained. The notion of rates of increase, embodying the essential ideas of the differential calculus, thus emerges.

Finally, theory and practice could be combined by finding graphs which approximately satisfy the simple functional laws which are being simultaneously studied.

I am quite aware that this suggestion of statistical study may seem fantastic, and perhaps be pronounced impossible. Of course, in a rapid first sketch, one cannot hope to have put all the details of the idea in their right relation to one another. But before the whole suggestion is definitely dismissed, I should like to know exactly why it is impossible and why it is fantastic. The information to be imparted is of the utmost importance for the subsequent conduct of life in self-governing communities; it illustrates important abstract ideas, and the means of study lie easily to hand. Also the course of work would involve simple definite efforts on the part of the students. This method of conducting the elementary study of mathematical analysis appears to me to be eminently practical, and at every stage to carry with it its own justification.

The mathematical treatment of our space-ideas is obviously of the first educational importance. I cannot pretend to be very satisfied with the immediate effects of the abolition of Euclid's Elements as a text-book. A lamentable deficiency in logical rigour has crept in, with entirely bad effects on the scholastic value of the subject. My belief is that the science as an educational instrument has been ruined from the time of Euclid downwards by fallacious views of logical method which seem to be both prevalent and traditional. There is an idea that the logical premises of a subject like geometry are propositions which have some peculiar quality of self-evidence, which is not merely one of degree. In fact, it is implied that there are natural premises which have to be used as such because they are self-evident and incapable of proof.

Probably, when the view is thus crudely stated it would be repudiated by everyone. But I believe it to be true that the usual presentation of geometry as a deductive science, based on axioms which the student is simply told to accept, does, in fact, habitually generate this fallacious notion; thereby, the harm done to a sound conception of the relation of logic to induction is nearly as great as is the good received from the training in the art of reason. The same error crops up in an even more pestilential form when authors on mechanics imply that that science is based on separate verifications of the various laws of motion. Half our difficulties in the elementary teaching of the deductive sciences arise from the tacit unconscious acceptance of this abominable heresy.

I am told that there are some animals whose centres of intelligence, such as they are, are fairly uniformly distributed throughout their bodies, so that, however you cut them in half, both parts are equally sensible. Something like this is the case in any science. The propositions which, for some reason or other, claim our credence, are distributed throughout the whole body of the subject. The function of deductive logic is, by the creation of a coherent logical system, to tie them together, so as to enable us to pool their evidence. But often there is more evidence for the more complex propositions than for the premises. The chief requisite for a premise is not obvious truth, but simplicity. There is no obvious truth about the law of gravitation; but the science of attractions, which is founded on it, is verified all along the line in so far as it is applied to ordinary matter.

Thus, in order to pool our evidence for a body of propositions to the utmost extent, it is desirable that the premises assumed should be as few and simple as possible, and, of course, the more fully they claim our credence the better. But none of these requisites are absolutely necessary on pain of logical fallacy. Our selection of premises is arbitrary, and must be guided by the purpose which we have in view. Now these logical considerations have a profound influence on our conceptions of the true mode in which to present geometry. They lead to the conclusion that the old traditional presentation is wrong.

In the first place it is of great importance that students, before considering any logical proofs, should be made thoroughly familiar with the set of ideas and propositions which are to compose the schedule of the subject. They should note that some propositions appear obvious, and that others are capable of experimental verification by the measurement of accurately drawn diagrams.

In this way, a schedule of important propositions should be thoroughly appreciated. Then a selection of some of the simpler and more obvious propositions should be made and treated as logical premises. They are *our* axioms of geometry, not *the* axioms of geometry; and from them, by the most rigid reasoning, justifying every step as we go along, the remainder of the schedule should be proved.

But I would remind you that there is no logical fallacy in retaining a logically redundant set of premises. We can, if we like, point out to our pupils that some of the assumed propositions can be proved from the others, but there is absolutely no necessity to give the proof if we think it too long or too difficult. What is necessary for education is that the pupil should definitely know what propositions have been assumed, that these axioms should carry with them some strong evidence of truth, and that the reasoning should be rigidly accurate and full. But where assumption is so easy, logical fallacies are unpardonable. Also it may be desirable to retain some propositions in the schedule for experimental consideration which are not ultimately subjected to logical proof. For example, the theory of similarity is the foundation of all maps and plans, and it is highly desirable to appreciate its elementary propositions even if they cannot be proved.

But, as in the case of algebra, the schedule should be rigidly purged of all propositions which might appear to the student to be merely curiosities without any important bearings.

But what are the important bearings of geometrical truths? In a sense, the science is its own justification. It is the framework almost instinctively adopted to state our experiences of the universe. In order to explain why we feel tired, we state the number of miles which we have walked; to explain why it took so many days to plough a field, we state its number of acres. In every attempted explanation of the material facts of life we have recourse to geometrical ideas. Geometry is the queen of physical sciences. Accordingly, in a sense, we might bring any geometrical theorem into the schedule. But our time is limited, and we shall do well to concentrate on a few truths of the widest application and of most immediate importance. Whatever we put into the schedule necessarily excludes something else, and this consideration governs our selection.

The treatment of the whole doctrine of similarity makes almost a small subject in itself. It faces us in the selection of the scales of our graphs and other diagrams. Also it naturally coalesces with the doctrine of arithmetical proportion which in its elements receives a simple algebraic treatment. This, again, finds an application in the proportional variation of the entries in statistical tables, corresponding to variations either of population or of other fundamental aggregates. I should like, at this point, to enter a plea for the inclusion of the parallelogram of forces and the polygon of forces as the fundamental example of the application of geometry to science. But already I have been led into a discourse which, against my original intention, has wandered away into a technical discussion.

We have been considering the place of elementary mathematics in a liberal education. What, in a few words, is the final outcome of our thoughts? It is that the elements of mathematics should be treated as the study of a set of fundamental ideas, the importance of which the student can immediately appreciate: that every proposition and method which cannot pass this test, however important for a more advanced study, should be ruthlessly cut out: that with the time thus gained, the fundamental ideas placed before the pupils can be considerably enlarged so as to include what in essence is the method of co-ordinate geometry, the fundamental idea of the differential calculus in relation to rates of increase, and the geometrical notion of similarity. Also, lastly, it has been insisted that important systematic applications of these ideas to the concrete world should be simultaneously studied—for example, some sets of social or scientific statistics and the use of the polygon of forces in the graphical solution of mechanical problems. Again, this rough summary can be further abbreviated into one essential principle, namely, simplify the details and emphasize the important principles and applications.

The suggestions which I have ventured to put forward have been made with unfeigned diffidence. I am emboldened to speak by the conviction that we have now a golden opportunity for reconstituting our scheme of mathematical education. But such opportunities are dangerous. If mathematical teaching is not now revivified by a breath of reality, we cannot hope that it will survive as an important element in the liberal education of the future.

Science in General Education

WE ARE BECOMING aware that in adjusting a curriculum, it is not sufficient to agree that some specified subject should be taught. We have to ask many questions and to make many experiments before we can determine its best relation to the whole body of educational influences which are to mould the pupil.

In the first place it is necessary to keep before our minds that nine-tenths of the pupil's time is, and must be, occupied in the apprehension of a succession of details—it may be facts of history, it may be the translation of a definite paragraph of Thucydides, it may be the observable effects in some definite physical experiment. You cannot learn Science, *passim;* what you do learn in some definite hour of work is perhaps the effect on the temperature of a given weight of boiling water obtained by dropping into it a given weight of lead at another definite temperature, or some analogous detailed set of facts. It is true that all teaching has its rhetorical moments when attention is directed to æsthetic values or to momentous issues. But practical schoolmasters will tell you that the main structure of successful education is formed out of the accurate accomplishment of a succession of detailed tasks. It is necessary to enforce this point at the very beginning of discussion, and to keep it in mind throughout, because the enthusiasm of reformers so naturally dwells on what we may term "the rhetoric of education."

Our second step in thought must be to envisage the principles which should govern the arrangement of the detailed lessons in the subject. An educational cynic will tell you that it does not make much difference what you teach the pupils: they are bound to forget it all when they leave school; the one important thing is, to get the children into the habit of concentrating their thoughts, of applying their minds to definite tasks, and of doing what they are told. In fact, according to this school of thought, discipline, mental and physical, is the final benefit of education, and the content of the ideas is practically valueless. An exception is made for pupils of unusual ability or of unusual twist of interest. I conceive this summary solution of the educational problem to be based on an entirely false psychology, and to be in disagreement with experience. It depends for its plausibility on the erroneous analogy of the intellectual organism with some kind of mechanical instrument such as a knife, which you first sharpen on a hard stone, and then set to cut a number of different things quite disconnected with the stone and the process of sharpening. The other sources of the theory are the disillusionment of tired teachers, and the trenchant judgments of those who will not give the time to think out a complex question. But as this opinion is not likely to be largely represented among members of the Congress, further contemplation of it is unnecessary. In considering the general principles which are to govern our selection of details, we must remember that we are concerned with general education. Accordingly we must be careful to avoid conceiving science either in quantity or quality as it would be presented to the specialists in that subject. We must not assume ample time or unusual scientific ability. Also in recent years the congestion of subjects in the curriculum, combined with the opposing claims of specialism, has led practically all English Schools and the Board of Education to adopt certain principles regulating the relations between general education and special subjects. Our discussion must take these for granted, if we wish to be practical. Education up to the age of sixteen, or sixteen-and-a-half, is to be dominated by the claims of general education, and extended attention to any special subject is to be limited by the claims of the whole balanced curriculum. In the case of a pupil of any reasonable ability there will be time for some specialism; but the ruling principle is, that where the claims of the two clash, the specialism is to be sacrificed to the general education. But after the age of sixteen, the position is reversed. The pupil is expected to devote the larger proportion of his time to some adequate special subject, such as classics, science, mathematics, or history, and the remaining portion to suitably contrasted subsidiary subjects, such as modern languages for a scientist or a mathematician. In other words, before sixteen the special subject is subsidiary to the general education, and after sixteen the general education is subsidiary to the special subject. Accordingly our discussion divides into two sections, namely, science in general education before the age of sixteen, and after the age of sixteen. The second division may also be taken to cover the University stage. This principle of a preliminary general education has set to educationalists a new problem which has not as yet been adequately worked on in any subject. Indeed it is only just dawning on responsible people in its full urgency. But on its solution depends the success of that modern system of education to which we are now committed.

The problem is this: In all schools, with negligible exceptions, the general education has to be arranged with practical uniformity for the school as a whole. In the first place it is not very certain who among the pupils are the future scientists, who the future classical scholars, or who are the future historians. For the greater number, the desirable differentiation will only gradually disclose itself. Secondly, we may not assume that the majority of boys or girls in secondary schools will remain at school after the age of seventeen, and thus continue any portion of the general education after the first period. Accordingly for both these reasons, the preliminary general training in each subject should form a self-contained course, finding its justification in what it has done for the pupil at its termination. If it is not justified then, it never will be, since at this point, in the vast majority of instances, the formal study of the subject ends.

If we examine the cause of the educational dissatisfaction at the end of the last and at the beginning of this century, we shall find that it centres round the fact that the subjects in the curriculum were taught as incomplete fragments. The children were taught their elementary mathematics exactly as though they were to proceed in later years to take their degrees as high wranglers. Of course most of them collapsed at the first stage; and nobody—least of all the children—knew why they had been taught just that selection of meaningless elaborate preliminaries. Anyhow, as they soon forgot it all, it did not seem much to matter. The same criticism applied to the classics, and to other subjects. Accordingly, every subject in the preliminary training must be so conceived and shaped as yielding, during that period, general aptitudes, general ideas, and knowledge of special facts, which, taken in conjunction, form a body of acquirement essential to educated people. Furthermore it must be shown that the valuable part of that body of acquirement could not be more easily and quickly gained in some other way, by some other combination of subjects.

In considering the framing of a scientific curriculum subject to these conditions, we must beware of the fallacy of the soft option. It is this pitfall which has ruined so many promising schemes of reform. It seems such an easy solution, that, in order to gain time, we should shape a course comprising merely the interesting descriptive facts of the subject and the more important and exciting generalizations. In this way our course is self-contained and can easily be compressed into a reasonable time. It will certainly be a failure, and the reason of the failure illustrates the difficulty or the art of education. In order to explain this, let us recur to the educational cynic whom I introduced at the beginning of this paper; for he really is a formidable critic. He will point out that in a few years your pupil will have forgotten the precise nature of any facts which you teach him, and will almost certainly have muddled your generalizations into incorrect forms. The cynic will ask, What is the use of a vague remembrance of the wrong date for the last glacial epoch, and of a totally erroneous idea of the meaning of "the survival of the fittest"? Furthermore, we may well doubt whether your science, as thus taught, will be really interesting. Interest depends upon background, that is to say, upon the relations of the new element of thought or perception to the pre-existing mental furniture. If your children have not got the right background, even "the survival of the fittest" will fail to enthuse them. The interest of a sweeping generalization is the interest of a broad high road to men who know what travel is; and the pleasure of the road has its roots in the labour of the journey. Again facts are exciting to the imagination in so far as they illuminate some scheme of thought, perhaps only dimly discerned or realized, some day-dreams begotten by old racial experience, or some clear-cut theory exactly comprehended. The complex of both factors of interest satisfies the cravings inherent in that mysterious reaching out of experience from sensation to knowledge, and from blind instinct to thoughtful purpose.

The conclusion is that you can only elicit sustained interest from a process of instruction which sets before the pupils definite tasks which keep their minds at stretch in determining facts, in illustrating these facts by ideas, and in illustrating ideas by their application to complex facts. I am simply enforcing the truism that no reform in education can abolish the necessity for hard work and exact knowledge.

Every subject in the general education must pull its weight in contributing to the building up of the disciplined power of definitely controlled thought. Experience amply proves that no one special traning is adequate for this purpose; the classical scholar cannot necessarily focus mathematical ideas, and the mathematician may be a slovenly thinker outside his science, and neither classic nor mathematician may have acquired the habits of procedure requisite for observation and analysis of natural phenomena. In this connection the function of the study of a subject is not so much to produce knowledge as to form habits. It is its business to transmute thought into an instinct which does not smother thought but directs it, to generate the feeling for the important sort of scientific ideas and for the important ways of scientific analysis, to implant the habit of seeking for causes and of classifying by similarities. Equally important is the habit of definitely controlled observation. It is the besetting fallacy of over-intellectual people to assume that education consists in training people in the abstract power of thought. What is important is the welding of thought to observation. The first effect of the union of thought and observation is to make observation exact. You cannot make an exact determination of the passing phenomena of experience unless you have predetermined what it is you are going to observe, so as to fix attention on just those elements of the perceptual field. It is this habit of predetermined perception and the instinctive recognition of its importance which is one of the greatest gifts of science to general education. It is here that practical work in the laboratory, or field work in noting geological or botanical characteristics, is so important. Such work must be made interesting to obtain the proper engrossment of attention, and it must be linked with general ideas and with adequate theory to train in the habit of pre-determining observation by thought. Every training impresses on its recipient a certain character; and the various elements in the general education must be so handled as to enrich the final character of the pupil by their contribution. We have been discussing the peculiar value of science in this respect. It should elicit the habit of first-hand observation, and should train the pupil to relate general ideas to immediate perceptions, and thereby obtain exactness of observation and fruitfulness of thought. I repeat that primarily this acquirement is not an access of knowledge but a modification of character by the impress of habit. Literary people have a way of relegating science to the category of useful knowledge, and of conceiving the impress on character as gained from literature alone. Accordingly I have emphasized this point.

We have, however, not yet exhausted the analysis of the impress on character due to science. The imagination is disciplined and strengthened. The process of thinking ahead of the phenomena is essentially a work of the imagination. Of course it involves only one specific type of imaginative functioning which is thus strengthened, just as poetic literature strengthens another specific type. Undoubtedly there will be some interplay between the types, but we must not conceive the imagination as a definite faculty which is strengthened as a whole by any particular imaginative act of a specific type. Accordingly science should give something to the imagination which cannot be otherwise obtained. If we are finally to sum up in one phrase the peculiar impress on character to be obtained from a scientific training, I would say that it is a certain type of instinctive direction in thought and observation of nature, and a facility of imagination in respect to the objects thus contemplated, issuing in a stimulus towards creativeness. We now turn to the other aspect of science. It is the systematization of supremely useful knowledge. In the modern world men and women must possess a necessary minimum of this knowledge, in an explicit form, and beyond this, their minds must be so trained that they can increase this knowledge as occasion demands. Accordingly the general education during the "pre-sixteen" period must include some descriptive summaries of physiological, botanical, physical, chemical, astronomical, and geological facts, even although it is not possible to choose all those sciences as subjects for serious study in the school curriculum. Especially this is important in the case of physiology owing to the accidental circumstance that we all have bodies.

We see therefore that the scientific curriculum must have a soft element and a hard element. The hard element will consist in the attainment of exact knowledge based on first-hand observation. The laboratory work will be so framed as to illustrate such concepts and theoretical generalizations as the pupil is to know. I would insist that science in this stage of education loses nearly all its value, if its concepts and generalizations are not illustrated and tested by practical work. This union of acquirement of concepts, of comprehension of general laws, of reasoning from them, and of testing by experiment will go slowly at first, because the child's powers of mind have to be built up. The pupil has not got the requisite generalizing faculty ready made, and it is the very purpose of the education to give it to him. Furthermore little bits of diverse sciences are useless for the purpose; with such excessive dispersion the systematic character of science is lost, nor does the knowledge go deep enough to be interesting. We must beware of presenting science as a set of pretentious names for obvious facts or as a set of verbal phrases. Accordingly the hard element in the scientific training should be confined to one or at most two sciences, for example, physics and chemistry. These sciences have also the advantage of being key sciences without which it is hardly possible to understand the others. By the age of sixteen every pupil should have done some hard work at these two sciences, and—generally speaking—it is scarcely possible that there will have been any time for analogous work in any other natural science, after the necessary mathematical time has been allotted. Probably in a four years' course the best quantitative division would be two years of physics and two years of chemistry, and mathematics all the time. But assuredly it is not desirable to do all the physics in the first period of two years, and all the chemistry in the second period. The first simple ideas clustering round the most elementary experiments will undoubtedly be physical and mechanical. But as some serious progress is made the two sciences illustrate each other, and also relieve each other by the width of interest thus developed. For example, the influence of physical conditions, such as temperature, on the rate, and even the possibility, of chemical transformations is an elementary lesson on the unity of nature more valuable than abstract formulation of statement on the subject.

Two factors should go to form the soft element in scientific education. The first and most important is browsing, with the very slightest external direction, and mainly dependent on the wayward impulses of a student's inward springs of interest. No scheme for education, and least of all for scientific education, can be complete without some facility and encouragement for browsing. The dangers of our modern efficient schemes remind one of Matthew Arnold's line[1] "For rigorous teachers seized my youth." Poor youth! Unless we are careful, we shall organize genius out of existence; and some measure of genius is the rightful inheritance of every man. Such browsing will normally take the form either of chemical experiments, or of field work in geology, or in zoology, or in botany, or of astronomical observation with a small telescope. Anyhow, if he can be got to do so, encourage the child to do something for himself according to his own fancy. Such work will reflect back interest on to the hard part of his training. Here the collector's instinct is the ally of science, as well as of art. Also it is surprising how many people—Shelley, for example—whose main interests are literary derive the keenest pleasure from divagations into some scientific pursuit. In his youth, the born poet often wavers between science and literature; and his choice is determined by the chance attraction of one or other of the alternative modes of expressing his imaginative joy in nature. It is essential to keep in mind, that science and poetry have the same root in human nature. Forgetfulness of this fact will ruin, and is ruining, our educational system. Efficient gentlemen are sitting on boards determining how best to adapt the curriculum to a uniform examination. Let them beware lest, proving themselves descendants of Wordsworth's bad man, they

[1] Stanzas from the *Grande Chartreuse*.

"Take the radiance from the clouds
In which the sun his setting shrouds."

The other factor should consist of descriptive lectures, designed for the purpose of giving necessary scientific information on subjects such as physiology, and also for the purpose of exciting general interest in the various sciences. No great amount of time need be taken up in this way. I am thinking of about three to six lectures a term. It should be possible to convey some arresting information about most sciences in this way, and in addition to concentrate on the necessary information on particular points which it is desired to emphasize. The difficulty about such lectures is that comparatively few people are able to give them successfully. It requires a peculiar knack. For this reason I suggest that there should be an exchange of lecturers between schools, and also that successful extension lecturers should be asked to take up this kind of work. It is evident that with a little organization and co-operation the thing could be done, though some care would be required in the arrangement of details. Finally we come to the position of science in general education after the age of sixteen. The pupil is now rapidly maturing and the problem assumes entirely a new aspect. We must remember that he is now engaged mainly in studying a special subject such as classics, or history, which he will continue during his subsequent University course. Among other things, his power of abstract thought is growing, and he is taking a keen delight in generalizations. I am thinking of boys in the sixth form and of undergraduates. I suggest that in general practical work should be dropped, so far as any official enforcement is concerned. What the pupil now wants is a series of lectures on some general aspects of sciences, for example, on the conservation of energy, on the theory of evolution and controversies connected with it, such as the inheritance of acquired characters, on the electromagnetic theory of matter and the constitutions of the molecule, and other analogous topics. Furthermore, the applications of science should not be neglected—machinery and its connection with the economic revolution at the beginning of the nineteenth century, the importance of nitrates and their artificial production, coal-tar, aeronautics, and other topics. As in the case of lectures at the earlier stage, not much time should be occupied by them, and also there is the same difficulty in finding the lecturers. I believe that these lectures are easier to give than the more elementary ones. But I think that it will still be found necessary to create some organization so that local talent can be supplemented by external aid.

Also at this stage books can be brought in to help; for example, Marett's *Anthropology* and Myres' *Dawn of History,* both in "The Home University Library," will form a bridge conducting the historians from the general theory of zoological evolution to the classical history which forms the commencement of their own special studies. I merely give this instance to show the sort of thing, and the scale of treatment, that I am thinking about. But this general treatment of science in the later stage of education will lose most of its value, if there is no sound basis laid in the education before the age of sixteen.

I will conclude with a general caution which summarizes the guiding principle of the preceding remarks: There is very little time, and so in the formal teaching above all things we must avoid both an aimless aggregation of details either in class or in laboratory and the enunciation of verbal statements which bring no concrete ideas to the minds of the pupils.

Historical Changes

THE EMANCIPATION OF WOMEN is inseparably connected with the development of social relations in this country. It is one of the great contributions of the American Republic to the life of mankind. England followed closely and supplied one great thinker who may rank as the intellectual founder of the modern phases of the movement. I mean John Stuart Mill, whose name should never be forgotten in these celebrations.

Those of us who, either by verbal tradition from a previous generation or by the relevant literature, can recall the prevalent tone of thought and of habit belonging to the earlier portion of the nineteenth century in Europe must do homage to the founders whom to-day we are honouring. They possessed courage and insight. They saw truly that the key of this great emancipation was education, and they acted with complete fearlessness.

Within this period change has not been confined to the emancipation of women. We live in a world of faster and faster transformation. An ancient sage has said, "No one crosses the same river twice." We can apply this saying to our own case: no one lectures to the same students twice; no one lectures on the same subject twice. The flux of the world has assumed a new relation to the spans and the period of human life.

As we think, we live. The mind is the crucible in which we fashion our purposes. The business of universities is the guidance of thought, its content of knowledge, its æsthetic apprehensions, and its activity of criticism.

We must not conceive the mentality of men as their private act of internal self-development. This private aspect of culture has been stressed far too strongly. The key to the history of mankind lies in this fact—as we think, we live.

Culture is the knowledge of the best that has been said and done, according to a famous definition of it. But such conceptions of culture, though true enough as far as they go, are defective. They are too static. They share the whole defect of the Renaissance movement upon which the ideals of the past four centuries have been founded. That movement conceived itself as the recovery of the models of a past civilization. It was based upon the notion of imitation.

Now there is great truth in this notion of culture. It always involves an imitation of the best that has been said and done. Yet something essential has been omitted in this characterization. That "something" is the profound flux of the world.

When knowledge of the distant past was more dim and the pace of change was slower, it was permissible to conceive the flux of the world as a turbulence of details amid an overpowering identity of principles. The changes were minor, the permanences were major.

To-day, this balance as between change and permanence has been decisively altered in two ways.

First, on a grand scale our cosmology discloses a process of overpowering change, from nebulæ to stars, from stars to planets, from inorganic matter to life, from life to reason and moral responsibility. We can no longer conceive of existence under the metaphor of a permanent depth of ocean with its surface faintly troubled by transient waves. There is an urge in things which carries the world far beyond its ancient conditions.

Secondly, on the small scale of the individual lives of men, the change in the conditions of social existence is recognizable within the life of one human being and almost within the span of one year.

It is natural at this point to remember Professor Lowes' analysis of Coleridge's poetic conception of life in Xanadu. In that ideal country, it seems that hopes and fears and actions were greatly influenced by "ancestral voices prophesying." Now I suggest to you that to-day in America "ancestral voices prophesying" are somewhat irrelevant. And for this reason: they do not know what they are talking about. The fact is that our honoured ancestors were largely ignorant of modern conditions, and so their prophecies are impressive, vaguely disturbing, but very unpractical.

I have been placing in sharp contrast two antithetical truths, one that culture is assimilation and imitation of what is best in the past, and the other that the transience of conditions renders the details of the past irrelevant to the present. The problem of modern education is contained in this antithesis.

It is the problem of the understanding of "history" in the greatest sense of that word. In so far as we fail in the education of youth—and of course we do fail— it is because we have not implanted in our students a right conception of their relation to their inheritance from the past. Almost all intellectual knowledge is derived from the past; our mental outfit consists of "ancestral voices prophesying." The criticism of knowledge is the criticism of the past. Whatever be the subject which we teach, our main task is to inculcate how to inherit, appreciatively and critically. What our students should learn is how to face the future with the aid of the past.

Knowledge is the reminiscence by the individual of the experience of the race. But reminiscence is never simple reproduction. The present reacts upon the past. It selects, it emphasizes, it adds. The additions are the new ideas by means of which the life of the present reflects itself upon the past.

Thus culture, besides involving a criticism of tradition, also requires a critical appreciation of novelty. A sane culture is not chiefly concerned with true or false, right or wrong, acceptance or rejection. These are crude extremes betokening a poor appreciation of the complexity of the world.

A new idea has its origin in explicit consciousness by reason of some relevance to the immediate situation. The first task is to appreciate the reason for its origin. What are the factors, logical, emotional, purposeful, or of direct novel perception, which have led to its appearance and its prevalence?

The next task is to define the proper importance of the novelty, to fix its status in the system of thought, and to determine its applications and its limitations in the sphere of action. We have to reduce the idea to its true proportions, and at the same time to express its importance within those proportions.

In respect to our reactions to novelty we are still living in the ancient Ages of Faith. "What went ye out into the wilderness to see? A reed shaken by the wind?"

Thanks to the labours of the eighteenth century, we have inherited an efficient system for the criticism of traditional thought. But in regard to novelty our critical apparatus is only half developed. Each generation runs into childish extremes. To-day we adore, and to-morrow we will flog, the images of our saints or at least desert their shrines.

This defect in our culture will never be remedied till we have discovered how to make the great secret of history effective in our way of understanding things. As yet we, who teach, cannot do it. This secret in the history of man is that every idea once was new, and for that reason was then vague, ill-defined, with glorious possibilities or with hideous consequences.

That "two and two make four" was once new and too abstract for importance. That "Cæsar should be murdered" was once a secret conjecture, and that "he had been murdered" was once a rumour. We treat the past merely as material for dissection, something settled and obvious, and we have no intimate feeling for the wavering steps of its advance. That "Cæsar is murdered" becomes merely an item in the abstract analysis of abstract history. Until mankind understands its own history, intimately as a concrete passage into an unknown future, our culture will never be adequate. We treat our novelties of to-day as though it were a novel fact that there should be novelty.

History is the drama of effort. The full understanding of it requires an insight into human toiling after its aim. In the absence of some common direction of aim adequately magnificent, there can be no history. The spectacle is then mere chaos.

The drama consists in the mixture of happiness and despair, of failure and victory, arising from the development of human purposes. It includes a tragedy and a comedy. But, as the Athenians well knew, no one is prepared for the relief of comedy until his passions have been purified by the tragic intensity. Comedy is the back-lash of tragedy, making life possible.

The drama of history is more than humour. It discloses an ultimate character in the nature of things, effecting a discrimination of human effort.

It is an easy sophism to dismiss the whole topic of this discussion with the saying that we should concentrate on the future and not on the past; that we want a forward-looking population. This is certainly true. But we cannot get rid of the past quite so easily. For if the past be irrelevant to the present, then the present and the understanding of it go together with the relevance of present to future. It is the business of a sound education to strengthen this sense of derivations and of consequences, and to provide it with understanding.

A weak spot in educational methods is here touched upon. We want an historical background, and even history itself fails to provide it in the required way. We want to get at the facts in the concrete, with their massive background of immediate life. History is apt to present us with the facts in the abstract, detached incidental curiosities. For example, mere lists of presidents of the United States and of Roman emperors are facts in the abstract.

Now every subject of study should be presented as in the abstract and in the concrete. Both sides are wanted. We learn them in the abstract, we feel them in the concrete. Every incident calls a halt in the flux of the world for the sake of its own massive immediate enjoyment. At the same time, it is to be conceived as a moment in the transition of form out of the past into the future.

For example, consider the jubilee which we are celebrating. Fifty years ago women did not go to college, to-day they do. This is a fact in the abstract, capable of clear statement in a short sentence. But the understanding of the difference in human life, now and fifty years ago, involved in this statement is the comprehension of the fact in the concrete. Still more concrete is the grasp of the mixture of waywardness and inevitableness with which the transition developed—an inevitableness which yet requires the commanding figures round which the drama revolves. The reason of all such celebrations, is the desire to make the past live, to turn abstract knowledge into the concrete feeling.

The function of art is to turn the abstract into the concrete and the concrete into the abstract. It elicits the abstract form from the concrete marble. Education, in every branch of study and in every lecture, is an art. The emphasis may be more on the abstract or more on the concrete. But always there remains the inescapable problem of marriage of form to matter.

Life is short and Art is long. We all fail in our efforts to present the essentials of culture to our students. It remains for their genius to convert our failure into success.

I discern decisive signs of the coming of a new epoch in American thought. The iconoclastic impulse which is so prominent in the literary school to-day has done its work. It is not rejected. It is not shocking anybody. But its preoccupations have ceased to interest the creative ability under thirty, still more that under twenty-five years of age. The struggle of elderly propriety with mid-

dle-aged destructive vehemence is an amusing spectacle
to the young. But it has no message for them: it stirs
them with no trumpet call.

Their interest is more directly æsthetic and construc-
tive. They are concerned with the beauty derived from
artistic finish of workmanship, with style, with restraint,
with balance. They seek the play of rapiers, in preference to the blows of sledge hammers.

But they are not mainly critical. Their criticism is
a subsidiary moment in their passage towards construction. Their effort after style is also, in like manner, subsidiary. They want to build an edifice of thought which
shall also be an edifice of beauty. Every variety of beauty
claims their interest—the logical beauty of scientific
thought, the beauty to be perceived by the senses, and
the beauty of conduct.

In one word, as you will already have seen, the young
of to-day are Athenian, in a sense in which no one belonging to the nineteenth century, either in England
or America, was Athenian. In this characterization I
am referring, of course, to the few, and not to the many.
Perhaps the movement will never attain to widespread
influence. It may fail in the luck of throwing up one or
two personalities of commanding power. But the trend
of interest is certainly there.

You will not misunderstand me. Among the young
people of to-day there is no one individual who satisfies,
even approximately, the many-sided Athenian ideal
which I have sketched. But, as you will remember, in
Athens itself there was no person who rose to the full
ideal of the typical Athenian. Plato knew well that the
ideal of a type is never incarnate in this dusty world.
Here again we meet, in a wider sense, the notion of
"imitation." It was Plato's phrase for the aim of individuals at the perfection of their type. We have already
encountered it as implicit in Arnold's conception of
culture.

In one sense an inflexible determinism reigns in this
world. For the making of an epoch is already settled

by the ideal which its youth set before themselves for imitation. As we think, we live.

In the shaping of this ideal, past and future fuse together in the present. The past is there as an inescapable fact, with its secret impress of modes of operation. In order to conjecture the boundary of possibility we must scan the past.

The pathway of mankind through history has been made visible to our understanding, in fact and in allegory, by that stream of immigrants who in ships across the ocean and in covered wagons across the prairies pursued the lure of their hopes to enlarge the boundaries of life.

They toiled forward, enjoying the stretch of their faculties, hunting, ploughing, starving, thirsting, dying. In this greatest story of the human race the heroism of women attained its utmost height.

Harvard: The Future

ABOUT TWENTY-FIVE YEARS for a man and about three hundred years for a university are the periods required for the attainment of mature stature. The history of Harvard is no longer to be construed primarily in terms of growth, but in terms of effectiveness.

I am talking of effectiveness in the wide world, of impress on the course of events, without which civilized humanity would not be as in fact it is. In the Cambridge of England, the first college was founded in the year 1284, and Emmanuel College in the year 1584. The English university was then grown up. Within the next one hundred and fifty years there occurred a brilliant period—*the* brilliant period—of European civilization. It staged a decisive episode in the drama of human life. In this episode the English university played no mean part, from Edmund Spenser and Francis Bacon at the outset to Newton and Dryden at the close. Among other things, Cambridge helped to contribute Milton, Cromwell, and Harvard University.

The term "European civilization" is now a misnomer, for the centre of gravity has shifted. Civilization haunts the borders of waterways. The shores of the Mediterranean and the western coasts of Europe are cases in point. But nowadays, relatively to our capacities, the dimensions of the world have shrunk, and the Atlantic Ocean plays the same rôle as the European seas in the former centuries. The total result is that the North American shores of the Atlantic are in the central position to influence the adventures of mankind, from East to West and from North to South. The static aspects of things are measured from the meridian of Greenwich but the world will rotate around the long line of American shores.

What is the influence of Harvard to mean in the immediate future, originating thought and feeling during the next fifty years, or during the next one hundred and fifty years? Harvard is one of the outstanding universities in the very centre of human activity. At the present moment it is magnificently equipped. It has enjoyed nigh seventy years of splendid management. A new epoch is opening in the world. There are new potentialities, new hopes, new fears. The old scales of relative quantitative importance have been inverted. New qualitative experiences are developing. And yet, beneath all the excitement of novelty, with its discard and rejection, the basic motives for human action remain, the old facts of human nature clothed in a novelty of detail. What is the task before Harvard?

It will be evident that in this summary presentation of the cultural problem the word "Harvard" is to be taken partly in its precise designation of a particular institution and partly as a symbolic reference to the university system throughout the Eastern states of this country. A closely intertwined group of institutions, the outcome of analogous impulses, has in the last three hundred years gradually developed, from Charlottesville to Baltimore, from Baltimore to Boston, and from Boston to Chicago. Of these institutions some are larger and some are smaller, some are in cities and some are in country places, some are older and some are younger. But each of them has the age of the group, as moulded by this cultural impulse. The fate of the intellectual civilization of the world is to-day in the hands of this group—for such time as it can effectively retain the sceptre. And to-day there is no rival. The Ægean coast line had its chance and made use of it; Italy had its chance and made use of it; France, England, Germany, had their chance and made use of it. To-day the Eastern

American states have their chance. What use will they make of it? The question has two answers. Once Babylon had its chance, and produced the Tower of Babel. The University of Paris fashioned the intellect of the Middle Ages. Will Harvard fashion the intellect of the twentieth century?

II

We cannot usefully discuss the organization of universities, considered as educational institutions, apart from a preliminary survey of the general character of human knowledge, and of some special features of modern life. Such a survey elicits perplexities which have troubled learning from the earliest days of the Greeks to the present moment. By introducing implicit assumptions in respect to these problems, it is possible to arrive at almost any doctrine respecting university organization.

In the first place, there is the division into certainty and probability. Some items we are certain about, others are matters of opinion. There is an obvious common sense about this doctrine, and its enunciation goes back to Plato. The class of certainties falls into two subdivisions. In one subdivision are certain large general truths—for example, the multiplication table, axioms as to quantitative "more or less"—and certain æsthetic and moral presuppositions. In the other subdivision are momentary discriminations of one's own state of mind: for example, a state of feeling—happiness at this moment; and for another example, an item of sense perception—that coloured shape experienced at this moment. But recollection and interpretation are both deceitful. Thus this latter subdivision just touches certainty and then loses it. There is mere imitation of certainty.

In the class of probabilities there are to be found all our judgments as to the goings on of this world of temporal succession, except so far as these happenings are qualified by the certainties whenever they are relevant.

I repeat my affirmation that, in some sense or other, this characterization of human knowledge is indubitable. No one doubts the multiplication table; also everyone admits that a witness on the witness stand can only produce fallible evidence, which the judicial authorities endeavour to assess, again only fallibly.

The bearing of these doctrines on the procedures of education cannot be missed. In the first place: Develop intellectual activities by a knowledge of the certain truths, so far as they are largely applicable to human life. In the second place: Train the understanding of each student to assess probable knowledge in respect to those types of occurrences which for any reason will be of major importance in the exercise of his activities. In the third place: Give him adequate knowledge of the possibilities of æsthetic and moral satisfactions which are open to a human being, under conditions relevant to his future life.

So far there is no disagreement. Unfortunately, exactly at this point our difficulties commence. This is the reason why the prefatory analysis was necessary. These difficulties are best explained by a slight reference to the history of thought, stretching from Greece to William James.

Plato was a voluminous writer, and apparently all his works have come down to us. They constitute a discussion of the various types of certain knowledge, of probable knowledge, and of æsthetic and moral ideals. This discussion, viewed as elucidating the above-mentioned classification of knowledge which is to be the basis of education, was a complete failure. He failed to make clear what was certain; and where he was certain, we disagree with him. He failed to make clear the relationship of things certain to things probable; and where he thought he was clear, we disagree with him. He failed to make clear the moral and æsthetic ends of life; and where he thought he was clear we disagree with him. No two of his dialogues are completely consistent with each other. No two modern scholars agree as to what any one dialogue exactly means. This failure of Plato is the great fact dominating the history of European thought.

Also this failure was typical. It stretches through every topic of human interest. Every single generalization respecting mathematical physics, which I was taught at the University of Cambridge during my student period from the years 1880 to 1885, has now been abandoned in the sense in which it was then held. The words are retained, but with different meanings.

The truth is that this beautiful subdivision of human knowledge, whether you make it twofold or threefold, goes up in smoke as soon as you try to fasten upon it any exact meaning. As a vague preliminary guide, it is useful. But when you trust it without reserve, it violates the conditions of human experience. The history of thought is largely concerned with the records of clear-headed men insisting that they at last have discovered some clear, adequately expressed, indubitable truths. If clear-headed men throughout the ages would only agree with each other, we might cease to be puzzled. Alas, that is a comfort denied to us.

III

The outcome of this brief survey is so fundamental in its relevance to education that it must be elucidated further by considering it in reference to two topics—Mathematics, and the Abiding Importance of Plato.

The science of Mathematics is the very citadel of the doctrine of certainty. It is unnecessary to bring the large developments of the subject into this discussion. Let us consider the multiplication table. This table is concerned with simple interrelations of cardinal numbers, as for example, "Twice three" is "six." Nothing can be more certain. But a little question arises: What are cardinal numbers? There is no universally accepted answer to this question. In fact, it is the battle ground of a controversy. The innocent suggestions which occur to us are traps which lead us into self-contradictions or into other puzzles. The notion of number is obviously concerned with the concept of a class, or a group, of many things. It expresses the special sort of many-ness

in question. Unfortunately the notion of a class is beset
with ambiguities leading to logical traps. We then have
recourse to the fundamental notions of logic, and again
encounter a contest of dissentient opinions. Logic is the
chosen resort of clear-headed people, severally convinced
of the complete adequacy of their doctrines. It is such
a pity that they cannot agree with each other.

Analogous perplexities arise in respect to the fundamental notions of other mathematical topics: for example, the meaning of the notions of a point, of a line,
and of a straight line. There is great confidence and no
agreement.

Thus the palmary instances of human certainty, Logic
and Mathematics, have given way under the scrutiny
of two thousand years. To-day we have less apparent
ground for certainty than had Plato and Aristotle. The
natural rebound from this conclusion is scepticism. Trust
your reflexes, says the sceptic, and do not seek to understand. Your reflexes are the outcome of routine. Your
emotions are modes of reception of the process. There
is no understanding, because there is nothing to understand.

Complete scepticism involves an aroma of self-destruc-
tion. It seems as the negation of experience. It craves
for an elegy on the passing of rational knowledge—the
beautiful youth drowned in the Sea of Vacuity.

The large practical effect of scepticism is gross acquiescence in what is immediate and obvious. Postpone-
ment, subtle interweaving, delicacies of adjustment, wide
co-ordinations, moral restraint, the whole artistry of civilization, all presuppose understanding. And without understanding they are meaningless.

Thus, in practice, scepticism always means some
knowledge, but not too much. It is indeed evident that
our knowledge is limited. But the traditional scepticism
is a reaction against an imperfect view of human knowledge.

It is in respect to this limitation of knowledge that
the ancient division into certainties and probabilities is
so misleading. It suggests that we have a perfectly clear

indication of the items in question, and are either certain or uncertain as to the existence of some definite connection between them. For example, it presupposes that we have a perfectly clear indication of the numbers 2 and 3 and 6, and are either certain or uncertain as to whether twice three is six.

The fact is the other way round. We are very vague as to the meanings of 1, and 2, and 3, and 5, and 6. But we want to determine these meanings so as to preserve the relations, "six is one more than five" and "twice three is six." In other words, we are more clear as to the interrelations of the numbers than as to their separate individual characters. We use the interrelations as a step towards the determinations of the things related.

This is an instance of the general truth, that our progress in clarity of knowledge is primarily from the composition to its ingredients. The very meaning of the notion of definition is the use of composition for the purpose of indication.

The important characterization of knowledge is in respect to clarity and vagueness.

The reason for this dominance of vagueness and clarity in respect to the problem of knowledge is that the world is not made up of independent things, each completely determinate in abstraction from all the rest. Contrast is of the essence of character. In its happy instances contrast is harmony; in its unhappy instances contrast is confusion. Our experience is dominated by composite wholes, more or less clear in the focus, and more or less vague in the penumbra, and with the whole shading off into umbral darkness which is ignorance. But throughout the whole, alike in the focal regions, the penumbral regions, and the umbral regions, there is baffling mixture of clarity and vagueness.

The primary weapon is analysis. And analysis is the evocation of insight by the hypothetical suggestions of thought, and the evocation of thought by the activities of direct insight. In this process the composite whole, the interrelations, and the things related, concurrently emerge into clarity.

One of the most interesting facts in the psychology of young students at the present time is the abiding interest of the platonic writings. From the point of view of displaying the sharp distinction between the certainties and the opinions involved in human knowledge, Plato failed. But he gave an unrivalled display of the human mind in action, with its ferment of vague obviousness, of hypothetical formulation, of renewed insight, of discovery of relevant detail, of partial understanding, of final conclusion, with its disclosure of deeper problems as yet unsolved. There we find exposed to our view the problem of education as it should dominate a university. Knowledge is a process, adding content and control to the flux of experience. It is the function of a university to initiate its students in the exercise of this process of knowledge.

IV

The problem before Harvard is set by the termination of an epoch in European culture. For three centuries European learning has employed itself in a limited definite task. It was a necessary task and an important task. Scholars, in science and in literature, have been brilliantly successful. But they have finished that task—at least for the time, although every task is resumed after the lapse of some generations. However, for the moment, the trivialization of the traditional scholarship is the note of our civilization.

The fundamental presupposition behind learning has been that of the possession of clear ideas, as starting points for all expression and all theory. The problem has been to weave these ideas into compound structures, with the attributes either of truth, or of beauty, or of moral elevation. There was presumed to be no difficulty in framing sentences in which each word and each phrase had an exact meaning. The only topics for discussion were whether the sentence when framed was true or false, beautiful or ugly, moral or shocking. European learning was founded on the dictionary; and splendid dictionaries were produced. With the culmination of the dictionaries the epoch has ended. For this reason, all the dictionaries of all the languages have failed to provide for the expression of the full human experience.

The ultimate cause for this characteristic of European learning was that from the close of the dark ages civilization had been progressing with the gradual recovery of the subtle, many-sided literature of the old classical civilization. Thought then had the character of a recovery of the wide variety of meanings embedded in Greek and Hellenistic written literature. The result was that everything that a modern scholar thought could have been immediately understood by Thucydides, or Democritus, or Plato, or Aristotle, or Archimedes. Any one of these men would have understood Newton's Laws of Motion at a glance. These laws were a new structure of old ideas. Perhaps Aristotle would have shied at Newton's first law. But he would have understood it. Any one of these men would have understood the American Declaration of Independence. There is nothing in the Constitution of the United States to puzzle them. Perhaps the addition of these five sages to an august tribunal might even facilitate the elucidation of its applications.

The conception of mind and matter, of motion in space, of individual rights, of the rights of social groups —the world of tragedy, and of joy, and of heroism— was thoroughly familiar to the ancients, and its obvious interrelations were expressed in language, and discussed, and rediscussed. Throughout the last three or four centuries the notion of learning was the discussion of the ways of the world with the linguistic tools derived from the past. This procedure of learning was the basis of progress from the simplicities of the dark ages to the modern civilization.

For this reason a narrow convention as to learning, and as to the procedures of institutions connected with it, has developed. Tidiness, simplicity, clarity, exactness, are conceived as characteristics of the nature of things, as in human experience. It is presupposed that a university is engaged in imparting exact, clear knowledge.

Lawyers are apt to presuppose that legal documents have an exact meaning, even with the absence of commas.

Thus, to a really learned man, matter exists in test tubes, animals in cages, art in museums, religion in churches, knowledge in libraries.

It is easy to sneer. But there is a problem here—a very difficult problem; and the success of Harvard depends upon maintaining a proper interweaving of its intricacies. The development of learning, and the success of education, require selection. The human mind can only deal with limited topics, which exclude the vague immensity of nature. Thus the tradition of learning is the solid ground upon which the university must be founded, in respect to both sides of its activity—namely, the enlargement of knowledge and the training of youth.

The real problem is to adjust the activities of the learned institution so as to suffuse them with suggestiveness. Human nature loses its most precious quality when it is robbed of its sense of things beyond, unexplored and yet insistent. Mankind owes its progress beyond the iron limits of custom to the fact that, compared to the animals, men are amateurs. "You Greeks are always children" is the taunt from Learning to Suggestiveness.

Learning is sensible, straightforward, and clear, if only you keep at bay the suggestiveness of things. This clarity is delusive, and is shot through and through with controversy. The traditional attitude of scholars is to choose a side, and to keep the enemy at bay by exposing their errors. Of course, in the clash of doctrine we must base thoughts and actions on those modes of statement which seem to express the larger truth. But it is fatal to dismiss antagonistic doctrines, supported by any body of evidence, as simply wrong. Inconsistent truths—that is, truths in the sense of conformity to some evidence— are seed beds of suggestiveness. The progress which they suggest lies at the very root of knowledge. It is concerned with the recasting of the fundamental notions on which the structure is built. The suggestion does not primarily concern a new conclusion. Fundamental progress has to do with the reinterpretation of basic ideas.

At this point, the problem has only been half stated. Experience does not occur in the clothing of verbal phrases. It involves clashes of emotion, and unspoken revelations of the nature of things. Revelation is the primary characterization of the process of knowing. The traditional theory of education is to secure youth and its teachers from revelation. It is dangerous for youth, and confusing to teachers. It upsets the accepted co-ordinations of doctrine.

Revelation is the enlargement of clarity. It is not a deduction, though it may issue from a deduction. The dictionaries are very weak upon this point.

v

Without doubt, in its preliminary stages education is concerned with the introduction of order into the mind of the young child. Experience starts as a "blooming, buzzing confusion." Order introduces enlargement, significance, importance, delicacies of perception. For long years the major aspect of education is the reduction of confusion to order, and the provision of weapons for this purpose.

And yet, even at the beginning of school life, it has been found necessary to interfuse the introduction of order with the enjoyment of enterprise. The balance is difficult to hold. But it is well known that education as mere imposed order of "things known" is a failure. The initial stages of reading, writing, and arithmetic should be suffused with revelation.

At the other end of education, during the university period, there is undoubtedly the excitement of novel knowledge—volumes of words. But an inversion has entered upon the stage. The child has to be taught the words that correspond to the things; the senior at college has lost the things that correspond to the words. His mind is occupied by literary scenery; by doctrines derived from books; by experiments of a selected character, with selected materials, and such that irrelevancies are neglected. Even his games are organized. Novel impulse is frowned upon at the bridge table, on the football field, and on the river. No member of a crew is praised for the rugged individuality of his rowing.

The question is how to introduce the freedom of nature into the orderliness of knowledge. The ideal of universities, with staff and students shielded from the contemplation of the sporadic life around them, will produce a Byzantine civilization, surviving for a thousand years without producing any idea fundamentally new.

There is no one recipe. It is an obvious suggestion to collect an able, vigorous faculty and give it a free hand, with every encouragement. This principle of university management has been no news at Harvard since its foundation. Also the environment of New England facilitates its practice, by producing both the men and the requisite atmosphere. It is not as simple to follow this suggestion as it looks. For half a century, on both sides of the Atlantic, I have been concerned with appointments. Nothing is more difficult than to distinguish between a loud voice and vigour, or a flow of words and originality, or mental instability and genius; or a big book and fruitful learning. Also the work requires dependable men. But if you are swayed too heavily by this admirable excellence, you will gather a faculty which can be depended upon for being commonplace.

Curiously enough, the achievements of the faculty do not depend on the exact judiciousness of each appointment. In a vigorous society, ability, in the sense of capacity for high achievement, is fairly widespread. Undoubtedly it can only be ascribed to a minority; but this minority is larger than it is conventional to estimate. The real question is to transmute the potency for achievement. The instrument for this purpose is the stimulus of the atmosphere. In other words, we come back to suggestiveness.

Knowledge should never be familiar. It should always be contemplated either under the aspect of novel application, or under the aspect of scepticism as to the extent of its application, or under the aspect of development of its consequences, or under the aspect of eliciting the fundamental meanings which it presupposes, or under the aspect of a guide in the adventures of life, or under the aspect of the æsthetic of its interwoven relationships, or under the aspect of the miraculous history of its discovery. But no one should remain blankly content with the mere knowledge that "twice three is six"—apart from all suggestion of relevant activity.

What the faculty have to cultivate is activity in the presence of knowledge. What the students have to learn is activity in the presence of knowledge.

This discussion rejects the doctrine that students should first learn passively, and then, having learned, should apply knowledge. It is a psychological error. In the process of learning there should be present, in some sense or other, a subordinate activity of application. In fact, the applications are part of the knowledge. For the very meaning of the things known is wrapped up in their relationships beyond themselves. Thus unapplied knowledge is knowledge shorn of its meaning.

The careful shielding of a university from the activities of the world around is the best way to chill interest and to defeat progress. Celibacy does not suit a university. It must mate itself with action.

There again a problem arises. The mere scattered happenings of daily affairs are veiled from our analysis. So far as we can see, they are chance issues. The real stimulation arises from the discovery of co-ordinated theory illustrated in co-ordinated fact; and the further discovery that the fact stretches so far beyond the theory, disclosing affiliations undreamed of by learning.

VI

The picture of a university now forms itself before us. There is the central body of faculty and students, engaged in learning, elaborating, criticizing, and appreciating the varied structure of existing knowledge. This structure is supported by the orthodox literature, by orthodox expositions of theory, by orthodox speculation, and by orthodox experiments disclosing orthodox novelty.

This prevailing orthodoxy is as it should be. So far as this orthodox expression has been systematized for the successful evocation of types of æsthetic experience, and the successful indication of the structural inter-relations of experience, and the successful demonstration of that structure—so far as this is accomplished, there is truth. We have argued that there is an inherent vagueness in the meanings employed and in the conformities reached. Thus the word "orthodoxy" has been employed to denote the vague, imperfect rightness of our formularized knowledge at any moment. Our knowledge and our skills are limited, and in the nature of things there is infinitude ever pressing new details into some clarity of discrimination.

Because of this imperfection, learned orthodoxy does well to ally itself where reason is playing some part in determining the patterns of occurrence. Orthodoxy can provide the controlled experiment. But here we pass to that partial control where some relevance is secured, but no detail of happenings. Such contact is gained by the absorption into the university of those schools of vocational training for which systematized understanding has importance. These are the professional schools which should fuse closely with the more theoretical side of university work. At present, their chief examples are the schools of Law, Religion, Medicine, Business, Art, Education, Governmental Activities, Engineering. The essential character of these schools is that they study the control of the practice of life by the doctrines of orthodoxy.

The main advantage to a university of this fusion of vocational schools with the central core of theoretical consideration is the increase of suggestiveness. The orthodoxy of reigning theories is a constant menace. By fusion with the schools the area of useful suggestiveness is doubled. It now has two sources. There is the suggestiveness of the vagrant intellect as it contemplates the orthodox expositions and the orthodox types of experiment. This is the suggestiveness of learning. But there is another suggestiveness derived from brute fact. Lawyers are faced with brute fact fitting into no existing legal classification. Religious experiences retain an insistent individuality. Each patient is a unique fact for a doctor. Business requires for its understanding the whole complexity of human motives, and as yet has only been studied from the narrow ledge of economics. Also Art, Education, and Governmental Activities are gold mines of suggestion. It is midsummer madness on the part of universities to withdraw themselves from the closest contact with vocational practices.

Curiously, the withdrawal of universities from close association with the practice of life is modern. It culminated in the eighteenth and nineteenth centuries, and heralds the decay of a cultural epoch.

I am not talking of the theories that men may have held at any time as to university functions. The point is as to the closeness of the relationship of the universities to the life around them—a closeness so natural as hardly to enter consciousness. In the first place, the universities arose out of nature, and were not exotic constructions imposed from above. The Papacy found universities; it did not devise them. Second, in studying the past we must distinguish between social barriers, trade secrets, and cultural doctrines.

In ancient Greece, whatever occupied a free citizen was worth study. That is why Socrates made himself a nuisance by cross-questioning people in the market place. He discovered the vagueness on which we have been insisting. Many things were done by slaves according to traditional methods. Nobody thought of lightening their labour; first, because it did not matter, and second, because there was no foreknowledge of the penetrating possibilities of modern science. Thus slave labour was a matter of course, without interest. But this is a social barrier, and not a doctrine of cultural activity. In the same way for the serfs of the Middle Ages. But here we must never forget the Benedictine monasteries and the variety of activities housed therein. Also the divine Plato was interested in drinking parties, and in the dances suitable for old gentlemen.

In a modern university the natural place for Aristotle would be somewhere between the Medical School, the Biological Departments, and the School of Education. But as life went on he would have looked in elsewhere. As to Plato, his two longest discourses are on political theory, the longer of the two being intensely practical. Also he made two long and dangerous journeys to give practical advice to governing people. His immediate pupils imitated his example. The Washington "brain trust" is not an American invention.

In the many centuries between Greece and our own times, the direct interplay between universities and practical affairs has been continuous. Salerno, Bologna, Paris, Edinburgh, and the Oxford of Jowett at once come to mind. In fact, almost any university with any length of history before the eighteenth century tells the same tale. As to men, it suffices to mention Erasmus, Locke, and Newton, among a thousand others.

The gross misunderstanding on this point arises from obliviousness of the part played by the great religious institutions, especially in the Middle Ages. They were concerned with actions, emotion, and thought. They co-ordinated intimacies of human feeling. The men directing their activities permeated universities and active life, the same men passing from one to the other of the two spheres. The rapid penetration by the mendicant orders into universities illustrates this point. The survival power of the great religious confederations demonstrates some large conformity of their procedures to the structure of human experience.

For a thousand years the Catholic Church was the deepest influence in the seats of learning and in the social relations of mankind. The mediæval universities were in touch with the life around them with a direct intimacy denied to their modern descendants. Of course, a large recasting of thought and doctrine was required. The first result was the brilliance of the seventeenth century. But household renovations are dangerous. For universities, the final result has been their seclusion from the variety of human feeling. To-day the activities oi the mediæval churchmen are best represented by the whole bundle of vocational activities, including those oi the various churches. In modern life, men of science are the nearest analogues to the mediæval clergy.

The mediæval clergy and the cultural humanism of the Hellenic world survive. Science (the search for order realized in nature), Hellenism (the search for value realized in human nature), Religion (the search for value basic for all things), express three factors belonging to the perfection of human nature. They can be studied apart. But they must be lived together in the one life of the individual. Thus there is a tidal law in the emphasis of epochs. At low tide factors are studied primarily in isolation. There is progress with manageable problems. The issue is trivialization; for meaning evaporates.

Importance belongs to the one life of the one individual. This is the doctrine of the platonic soul. At the high tide, combinations of factors dawn on consciousness with the importance of vivid shadows of this full unity of experience. And the knowledge in the low tide has required the high tide to provide compositions as material for thought.

VII

A university should be, at one and the same time, local, national, and world-wide. It is of the essence of learning that it be world-wide, and effectiveness requires local and national adaptations. It is not easy to hold the balance. But unless this difficult balance be held with some genius, the university is to that extent defective.

New England provides the near environment for Harvard, and from that local environment the institution derives its marked individuality, which is its strength. Also the most direct mission for Harvard is to serve the whole of these United States. The maintenance of a great civilization on this continent, from ocean to ocean, is the first purpose of American university life.

But the ideal of the good life, which is civilization—the ideal of a university—is the discovery, the understanding, and the exposition of the possible harmony of diverse things, involving and exciting every mode of human experience. Thus it is the peculiar function of a university to be an agent of unification. This does not mean the suppression of all but one. With this ideal before it, the notion of bare suppression sends a shiver through the academic framework. It savours of treason. Even local limitations are but means to the highest of all ends. Even methods are limitations. The difficulty is to find a method for the transcendence of methods. The living spirit of a university should exhibit some approach to this transcendence of limits.

The pursuit of harmony has its difficulties, alike in the realm of action and in the realm of understanding and in the realm of æsthetic enjoyment. The ideal of final harmony lies beyond the reach of human beings. Thus any civilized culture exhibits a mixture of harmony and discord. The university is struggling with discord in its journey toward harmony. It is spreading the enjoyment of such harmonies as the human tradition at that moment conveys, and it is pioneering in the prairies of disordered experience.

When all has been said, the universe is without bounds, learning is world-wide, and the springs of emotion lie below conventionalities. You cannot limit the sources of a great civilization; nor can you assign the stretch of its influence.

To-day Harvard is the greatest of existing cultural institutions. The opportunity is analogous to that of Greece after Marathon, to that of Rome in the reign of Augustus, to that of Christian institutions amid the decay of civilization. Each of these examples recalls tragic failure. But in each there is success which has secured enrichment of human life. If Greece had never been, if Augustan Rome had never been, if Institutional Christianity had never been, if the University of Paris had never been, human life would now be functioning on a lower level, nearer to its animal origins. Will Harvard rise to its opportunity, and in the modern world repeat the brilliant leadership of mediæval Paris?

PART IV

SCIENCE

The First Physical Synthesis

THERE ARE IN THE history of civilization certain dates which stand out as marking either the boundaries or the culminations of critical epochs. It is true that no epoch either commences, ends, or sums itself up in one definite moment. It is brought upon the stage of reality in the arms of its predecessors, and only yields to its successor by reason of a slow process of transformation. Its terminals are conventional. Wherever you choose to fix them, you can be confronted with good reasons for an extension or contraction of your period. But the meridian culmination is sometimes unmistakable, and it is often marked by some striking events which lend an almost mystic symbolism to their exact date. Such a date is the year 1642 of our epoch, the year in which occurred the death of Galileo and the birth of Newton. This date marks the centre of that period of about 100 years during which the scientific intellect of Europe was framing that First Physical Synthesis which has remained down to our own times as the basis of science. The development of modern Europe from the world of the Renaissance and the Reformation is unintelligible in its unique importance without an understanding of the achievements of these two men. The great civilizations of Asia and of the classical times in the Mediterranean had their epochs of artistic and literary triumph, of religious reformation, and of active scientific speculation. But it was the fortune of modern Europe that during the seventeenth century, amid a ferment of scientific speculation, two men, one after the other, appeared, each with a supreme gift of physical intuition, with magnificent powers of abstract generalization, and each with subsidiary endowments exactly suited to the immediate circumstances of the scientific problem, this one a supreme experimentalist and enough of a mathematician, and that one a supreme mathematician and enough of an experimentalist. Archimedes left no successor. But our modern civilization is due to the fact that in the year when Galileo died, Newton was born. Think for a moment of the possible course of history supposing that the life's work of these two men were absent. At the commencement of the eighteenth century many curious and baffling facts of physical science would have been observed, vaguely connected by detached and obscure hypotheses. But in the absence of a clear physical synthesis, with its overwhelming success in the solution of problems which from the most remote antiquity had excited attention, the motive for the next advance would have been absent. All epochs pass, and the scientific ferment of the seventeenth century would have died down. Locke's philosophy would never have been written; and Voltaire when he visited England would have carried back to France merely a story of expanding commerce and of the political rivalries between aristocratic factions. Europe might then have lacked the French intellectual movement. But the Fates do not always offer the same gifts twice, and it is possible that the eighteenth century might then have prepared for the western races an intellectual sleep of a thousand years, prosperous with the quiet slow exploitation of the American continent, as manual labour slowly subdued its rivers, its forests, and its prairies. I am not concerned to deny that the result might have been happier, for the chariot of Phœbus is a dangerous vehicle. My only immediate thesis is that it would have been very different.

The forms of the great works by which the minds of

Galileo and Newton are best known to us bear plain evidence of the contrast between their situations. In his book entitled, *The Two Systems of the World in Four Dialogues,* and published in 1632, Galileo is arguing with the past; whilst in his *Mathematical Principles of Natural Philosophy,* published in 1687, Newton ignores old adversaries and discussions, and, looking wholly to the future, calmly enunciates definitions, principles, and proofs which have ever since formed the basis of physical science. Galileo represents the assault and Newton the victory. There can be no doubt but that Galileo is the better reading. It is a real flesh and blood document of human nature which has wedged itself between the two austere epochs of Aristotelian Logic and Applied Mathematics. It was paid for also in the heart's blood of the author.

The catastrophe happened in this way: most unfortunately His Holiness, the reigning pope, in an entirely friendly interview after the Inquisition had forbidden the expression of Copernican opinions, made use of the irrefutable argument that, God being omnipotent, it was as easy for him to send the sun and the planets round the earth as to send the earth and the planets round the sun. How unfortunate it is that even an infallible pontiff and the greatest of men of science, with the most earnest desire to understand each other, cannot rid themselves of their presuppositions. The pope was trembling on the verge of the enunciation of the relativity of motion and of space, and in his Dialogues there are passages in which Galileo plainly expresses that same doctrine. But neither of them was sufficiently aware of the full emphasis to be laid upon that truth. Accordingly the next precious ten minutes of the conversation in which Galileo might have cleared away the little misunderstanding were wasted, and as a result there ensued for the world's edification the persecution of Galileo and a clear illustration of the limits of infallibility. The true moral of the incident is the importance of great men keeping their tempers. Galileo was annoyed—and very naturally so, for it was an irritating sort of argument with which to counter a great and saving formulation of scientific ideas. Unfortunately he went away and put the pope's argument into the mouth of Simplicius, the man in the Dialogues who always advances the foolish objections. It is welcomed in the following speech by the leading interlocutor, Salviatus—I give it in the seventeenth-century translation of Thomas Salusbury:

"This of yours is admirable, and truly angelical doctrine, to which very exactly that other accords in like manner divine, which whilst it giveth us leave to dispute, touching the constitution of the world, addeth withall (perhaps to the end that the exercise of the minds of men might neither be discouraged nor made bold) that we cannot find out the works made by his hands. Let therefore the Disquisition permitted and ordained us by God, assist us in the knowing, and so much more admiring his greatness, by how much less we find ourselves too dull to penetrate the profound abysses of his infinite wisdom."

At this point the Dialogues end. Galileo always protested that he had meant no discourtesy. But the pope, even if his infallibility tottered, was here assisted by the gift of prophecy and smelt Voltaire. Anyhow in his turn he lost his temper and afterwards remained the bitter enemy of Galileo.

Galileo's supreme experimental genius is shown by the way in which every hint which reached him is turned to account and immediately made to be of importance. He hears of the telescope as a curiosity discovered by a Dutch optician. It might have remained a toy, but in his hands it created a revolution. He at once thought out the principles on which it was based, improved upon its design so as to obviate the inversion of objects, and immediately applied it to a systematic survey of the heavens. The results were startling. It was not a few details that were altered, but an almost sacred sentiment which fell before it. I have often thought that the calmness with which the Church accepted Copernicus and its savage hostility to Galileo can only be accounted for by measuring the ravages made by the telescope on the sacred doctrine of the heavens. It was then seen too late that the Copernican doctrine was the key to the position. But Galileo's Dialogues plainly show that it was not the movement of the earth but the glory of the heavens which was the point at issue. It must be remembered that the heaven, which Christ had taught is within us, was by the popular sentiment of mediæval times placed above us. Accordingly when the telescope revealed the moon and other planets reduced to the measure of the earth, and the sun with evanescent spots, the shock to sentiment was profound. It is the characteristic of shocked sentiment in the case of men whose learning surpasses their genius that they begin to quote Aristotle. Accordingly Aristotle was hurled at Galileo.

The Dialogues are the records of the contemporary dispute between Galileo and the current Aristotelian tradition, and the end of the discussion was the creation of the modern scientific outlook of which Galileo was the first perfect representative—somewhat choleric but entirely whole-hearted.

So far we have been endeavouring to appreciate the climate of opinions amid which Galileo's life was passed —and you will remember that no climate is composed of a succession of uniform days, especially in its spring-time. A judicious selection could affix almost any label to the thought of the seventeenth century. What we have to keep in our minds is that at its beginning, so far as science was concerned, men knew hardly more than Aristotle and less than Archimedes, while at its end the main positions of modern science were firmly established.

I will now endeavour to explain the main revolutionary ideas which Galileo impressed upon his contemporaries. The first one was the doctrine of the uniformity of the material universe. This doctrine is now so obvious to us that we can only think of it in the attenuated form of discussions on miracles or on the relations of mind and matter. But in Galileo's time the denial of uniformity went much deeper than that. The different regions of Nature were supposed to function in entirely different ways. This presupposition led to a style of argument which is foreign to our ears. For example, here is a short speech of Simplicius, the upholder of the old Aristotelian tradition in Galileo's Dialogues, chosen almost at random:—

> Aristotle, though of a very perspicacious wit, would not strain it further than needed: holding in all his argumentations, that sensible experiments were to be preferred before any reasons founded upon strength of wit, and said those which should deny the testimony of sense deserved to be punished with the loss of that sense; now who is so blind, that sees not the parts of the Earth and Water to move, as being grave, naturally downwards, namely, towards the centre of the Universe, assigned by nature herself for the end and term of right motion *deorsum;* and doth not likewise see the Fire and Air to move right upwards towards the Concave of the Lunar Orb, as to the natural end of motion *sursum?* And this being so manifestly seen, and we being certain, that *eadem est ratio totius et partium,* why may we not assert it for a true and manifest proposition, that the natural motion of the Earth is the right motion *ad medium,* and that of the Fire, the right *a medio?*

In this passage we note that different functions are assigned to the Centre of the Universe to which the Earth or any part of it naturally moves in a straight line, and to the Concave of the Lunar Orb (to which Fire naturally moves in a straight line). The idea of the neutrality of situation and the universality of physical laws, regulating casual occurrences and holding indifferently in every part, is entirely absent. On the contrary, each local part of nature has its one peculiar function in the scheme of things. It is a fine conception: the only objection to it is that it does not seem to be true. I am not sure, however, that the Einstein conception of the physical forces as being due to the contortions of space-time is not in some respects a return to it.

But let us see how Galileo in the person of the interlocutor, Salviatus, answers this speech of Simplicius. His answer is somewhat long, and I only give the relevant part:—

. . . Now, like as from the consentaneous conspiration of all the parts of the Earth to form its whole, doth follow, that they with equal inclination concur thither from all parts; and to unite themselves as much as is possible together, they there physically adapt themselves; why may we not believe that the Sun, Moon, and other mundane Bodies, be also of a round figure, not by other than a concordant instinct, and natural concourse of all the parts composing them? Of which, if any, at any time, by any violence were separated from the whole, is it not reasonable to think, that they would spontaneously and by natural instinct return? and in this manner to infer, that the right motion agreeth with all mundane bodies alike.

Note that in this answer Galileo, in the person of Salviatus, entirely ignores any peculiar function or property to be assigned to a Centre of the Universe or to a Concave of the Lunar Orb. He has in his mind the conceptions of modern science, in that the Earth, the Moon, the Sun, and the other planets are all bodies moving in an indifferent neutral space, and each attracting its own parts to form its whole—or, as Salviatus puts it, "the consentaneous conspiration of all the parts of the Earth to form its whole."

Evidently Galileo is very near to the Newtonian doctrine of Universal Gravitation. But he is not quite there. Newton enunciates the doctrine that every particle of matter attracts every other particle of matter in a certain definite way. Galileo—as children say in the game of Hide-and-Seek—is very hot in respect to this doctrine. But he does not seem, at least in this passage, to have made the final generalization. He is thinking particularly of the Earth, the Sun, the Moon, and other planets—and his guardian angel does not appear to have whispered to him the generalization "any material body." Newton probably knew Galileo's Dialogues nearly by heart. They were standard works in his time. Cannot we imagine him sitting in his rooms between the gateway and the chapel of Trinity College, or in the orchard watching the apple fall, and with this passage of Galileo's

Dialogue running in his mind, perhaps the very words of Salusbury's translation which I have quoted, "the consentaneous conspiration of all the parts of the Earth to form its whole." Suddenly the idea flashes on him— "What are the Earth and the Sun and the Moon? Why, they are any bodies! We should say therefore that any bodies attract. But if this be the case, the Earth and the Sun and the Moon attract each other, and we have the cause maintaining the planets in their orbits." In this course of thought Newton would have been assisted by his third law of motion. For by it if the Earth attracts the apple, then the apple attracts the Earth.

By this conjectural reconstruction of Newton's state of mind we see that, given a genius with adequate mathematical faculties, Newton's Principia is the next step in science after Galileo's Dialogues. Probably Galileo himself would have gone farther in this direction if his imagination had not been hampered by the necessity of arguing with the Conservative Party. It is in general a mistake to waste time in discussions with people who have the wrong ideas in their heads. But in Galileo's time and country the Conservative Party had thumb-screws at its service and could thereby enforce a certain amount of attention to its ideas.

Undoubtedly the whole implication of the answer of Salviatus is that the Earth, Sun, etcetera, are mere bits of matter. It is difficult for us to estimate how great an advance Galileo made in adumbrating this position. Consider, for example, this statement by Simplicius, made in another connection, enforcing a doctrine which he upholds throughout the whole of the Dialogues:—

See here for a beginning, two most convincing arguments to demonstrate the Earth to be most different from the Cælestial bodies. First, the bodies that are generable, corruptible, alterable, &c., are quite different from those that are ingenerable, incorruptible, unalterable, &c. But the Earth is generable, corruptible, alterable, &c., and the Cælestial bodies ingenerable, incorruptible, unalterable, &c. Therefore the Earth is quite different from the Cælestial bodies.

That is the sort of thing that Galileo was up against, not as a mere casual idea occurring to a subtle reasoner, but as the very texture of current notions. The primary achievement of the first physical synthesis was to clear all this away. Galileo with his telescope, his trenchant, bold intellect, and his magnificent physical intuition was the man who did it.

But we have not nearly exhausted Galileo's contributions to the general ideas of science. We owe to Galileo the First Law of Motion. Probably most of us have in our minds Newton's enunciation of this law, "Every body continues in its state of rest or of uniform motion in a straight line except so far as it is compelled by impressed force to change that state." This is the first article of the creed of science; and like the Church's creeds it is more than a mere statement of belief: it is a pæan of triumph over defeated heretics. It should be set to music and chanted in the halls of Universities. The defeated adversaries are the Aristotelians who for two thousand years imposed on Dynamics the search for a physical cause of motion, whereas the true doctrine conceives uniform motion in a straight line as a state in which every body will naturally continue except so far as it is compelled by impressed force to change that state. Accordingly in Dynamics we search for a cause of the change of motion, namely either a change in respect to speed or a change in respect to direction of motion. For example, an Aristotelian investigating the motion of the planets in their orbits would seek for tangential forces to keep the planets moving; but a follower of Galileo seeks for normal forces to deflect the direction of motion along the curved orbit. This is why Newton, at the moment which we pictured him as he sat in his rooms in Trinity College thinking about gravitation, at once saw that the attraction of the Sun was the required force. It was nearly normal to the orbits of the planets. Here again we see how immediately Newton's physical ideas follow from those of Galileo. One genius completes the work of the other.

It has been stated by Whewell that in his Dialogues

on the Two Principal Systems of the World Galileo does not enunciate the first law of motion, and that it only appears in his subsequent Dialogues on Mechanics. This may be formally true so far as a neat decisive statement is concerned. But in essence the first law of motion is presupposed in the argumentation of the earlier dialogues. The whole explanation why loose things are not left behind as the Earth moves depends upon it.

Galileo also prepared the way for Newton's final enunciation of the Laws of Motion by his masterly investigation of the uniform acceleration of falling bodies on the Earth's surface and his demonstration that this acceleration is independent of the relative weights of the bodies, except so far as extraneous retarding forces are concerned. He swept away the old classification of natural and violent motions as founded on trivial unessential differences, and left the way entirely open for Newton's final generalizations. Newton conceived explicitly the idea of a neutral absolute space within which all motion is to be construed, and of mass as a permanent intrinsic physical quantity associated with matter, unalterable except by the destruction of matter. He phrased this concept in the definition, mass is quantity of matter. He then conceived the true measure of force as being the product of the mass of the body into its rate of change of velocity. The importance of this conception lies in the fact that force as thus conceived is found to depend on simple physical conditions, such as mass, electric and magnetic charges, electric currents, and distances. We owe to Newton the final formulation of the basic physical ideas which have served science so well during these last two centuries. They comprise the foundations of the science of Dynamics, and Law of Gravitation. We also owe to Galileo's experimental genius the telescope and its first systematic use in science, the pendulum clock (subsequently perfected by Huyghens) and the experimental demonstration of the laws of falling bodies. To Newton's mathematical genius we owe the deduction of the properties of the planetary orbits from dynamical principles. To Galileo and Newton we must add the name of Kepler so far as astronomy is concerned, and of Stevinus of Bruges so far as mechanics is concerned. He discovered the famous triangle of forces. But in one lecture lasting one hour you will not expect me to give a detailed account of the science of the seventeenth century.

In like manner we must add the name of Huyghens in mentioning the services of Galileo and Newton to the science of Optics. Huyghens first suggested the undulatory theory of light, to be revived at the beginning of the nineteenth century by Thomas Young and Fresnel. But the immediately fruitful work was due to Galileo with his studies on the theory of the telescope, and to Newton with his studies on the theory of colour. Both Dynamics and Optics reached Galileo as a series of detached truths (or falsehoods) loosely connected. After the work of Galileo and Newton they emerged as well-knit sciences on firm foundations.

Galileo's preoccupation with Optics doubtless helped him to another great idea which has coloured all modern thought. Light is transmitted through space from its origin by paths which may be devious and broken. What you see depends on the light as it enters your eye. You may see a green leaf behind the looking-glass; but the leaf is really behind your head and you are really looking at its image in the mirror. Thus the green which you see is not the property of the leaf, but it is the result of the stimulation of the nerves of the retina by the light which enters the eye. These considerations led Descartes and Locke to elaborate the idea of external nature consisting of matter moving in space and with merely primary qualities. These primary qualities are its shape, its degree of hardness and cohesiveness, its massiveness, and its attractive effects and its resilience. Our perceptions of nature such as colour, sound, taste and smell, and sensations of heat and cold form the secondary qualities. These secondary qualities are merely mental projections which are the result of the stimulation of the brain by the appropriate nerves. Such in outline is the famous theory of primary and secondary qualities in the form in which it has held the field during the modern period of science. It has been of essential service in directing scientific investigation into fruitful fields both of physics and physiology. Now the credit for its first sketch is due to Galileo. Here is an extract from Galileo's work, *Il Saggiatore,* published in 1624. I take it from the English life of Galileo by J. J. Fahie:—

"I have now only to fulfill my promise of declaring my opinions on the proposition that motion is the cause of heat, and to explain in what manner it appears to me that it may be true. But I must first make some remarks on that which we call heat, since I strongly suspect that a notion of it prevails which is very remote from the truth; for it is believed that there is a true accident, affection, or quality, really inherent in the substance by which we feel ourselves heated. This much I have to say, that as soon as I form a conception of a material or corporeal substance, I simultaneously feel the necessity of conceiving that it has boundaries, and is of some shape or other; that relatively to others it is great or small; that it is in this or that place, in this or that time; that it is in motion or at rest; that it touches, or does not touch another body; that it is unique, rare, or common; nor can I, by any act of imagination, disjoin it from these qualities; but I do not find myself absolutely compelled to apprehend it as necessarily accompanied by such conditions as that it must be white or red, bitter or sweet, sonorous or silent, smelling sweetly or disagreeably; and if the senses had not pointed out these qualities, it is probable that language and imagination alone could never have arrived at them. Therefore I am inclined to think that these tastes, smells, colours, &c., with regard to the object in which they appear to reside, are nothing more than mere names, and exist only in the sensitive body; insomuch that when the living creature is removed, all these qualities are carried off and annihilated; although we have imposed particular names upon them (different from those other and real accidents), and would fain persuade ourselves that they truly and in fact exist. But I do not believe that there exists anything in external bodies for exciting tastes, smells and sounds, but size, shape, quantity, and motion, swift or slow; and if ears, tongues, and noses were removed, I am of opinion that shape, quantity, and motion would remain, but there would be an end of smells, tastes, and sounds, which, abstractedly from the living creature, I take to be mere words."

If we knew nothing else about Galileo except that in the October of the year 1623 he published this extract, we should know for certain that a man of the highest philosophic genius then existed. On the subject of this extract, he leaves nothing for Descartes and Locke to do, except to repeat his statement in their own language, and to emphasize its philosophic importance. Indeed in many ways this original statement by Galileo is, as I believe, more accurately and carefully drawn than the usual formulations of modern times which I followed in my introductory remark.

I will now quit the special consideration of Galileo and Newton. I hope that I have with sufficient clearness given my reasons for holding that they are to be considered as the parents of modern science and as the joint authors of the first physical synthesis. You cannot disentangle their work. There would have been no Newton without Galileo; and it is hardly a paradox to say, that there would have been no Galileo without Newton. Galileo was the Julius Cæsar and Newton the Augustus Cæsar of the empire of science.

But these men did not work in a vacuum. It was an age of ferment, and they had as contemporaries men with genius all but equal to theirs. Francis Bacon was a contemporary of Galileo, somewhat older (1561-1626). I need not remind you that Bacon was the apostle of the experimental method. He especially emphasized the importance of keeping our minds open throughout a careful and prolonged examination of the facts. Like all apostles he somewhat exaggerated his message, and perhaps undervalued the importance of provisional theories. But the main point is perfectly correct and particularly important in view of the tradition of the preceding 1500 years, during which experiment had languished. Aristotle had discovered the importance of classification, and neither he nor his followers had realized the danger of classification proceeding on slight and trivial grounds. The greatest curse to the progress of science is a hasty classification based on trivialities. An example of what I mean is Aristotle's classification of motions into violent and natural. Bacon's writings were a continual protest against this pitfall. Again the active life of Descartes lies between those of Galileo and Newton. He published his *Principia Philosophiae* in 1644, just two years after the date which I have assigned as the symbolic centre of the epoch. The general concepts of space and matter, body and spirit, as they have permeated the scientific world, are largely in accordance with the way in which he fashioned them. He viewed space as a property of matter and therefore rejected the idea of purely empty space. This conception of space as an essential plenum led him to speculate on the other physical characteristics of the stuff whose extension is space. He thus hit on the idea of the vortices which carry along the heavenly bodies. These vortices are a failure. For one thing, they show that Descartes had not really assimilated the full import of Galileo's work in his discovery of the first law of motion. The planets do not want anything to carry them along, and that is just what Descartes provides. But for all that I hold that Descartes with his plenum was groping towards a very important truth which I will endeavour to explain before I finish this lecture. Newton's formulation of gravitation led Newton's followers to insist on the possibility of a vacuum, but the nineteenth century again filled space with an ether. Finally Einstein has recurred to the inversion of Descartes' doctrine and has made matter a property of space. The Newtonian vacuum and the Cartesian plenum have fought a very equal duel during the last few centuries. Leibniz, Newton's contemporary, emphasized the relativity of space.

This mention of relativity leads me to my last topic, which is to ask, how to-day we would criticize this First

Physical Synthesis which we owe to the seventeenth century.

In the first place, if we are wise, before criticizing it we will stop to admire it, and to note its essential services to science, and (in its main outlines) its continuing value to-day. We must do honour to the century of genius to which we owe it—a century which will compare with the greatest that Greece can show.

By a criticism of the great physical synthesis which is the legacy of this century to science I do not mean a mere enumeration of the additions since made, for example, the rise of the concept of energy, of the atomic theory, or of the theory of various chemical elements. Such homogeneous additions leave the concept undisturbed. In this way, Kelvin made it the mainspring of all his scientific speculations. But for the last thirty years or so, the great ideas of the seventeenth century have, so to speak, been losing their dominating grip on physical science.

Clerk-Maxwell probably thought that he had finally established its ascendancy. In truth he had set going trains of thought which in the hands of his followers have caused it to totter. Galileo and his followers thought in terms of time, space, and matter. They were in fact more Aristotelian than they knew—though they wore their Aristotle with a difference. Clerk-Maxwell emphasized the importance of the electromagnetic field as an interplay of relations between various electromagnetic quantities. Maxwell himself looked on this field as merely expressing strains, stresses, and motions of the ether, a point of view quite in the Galilean tradition. But recently the field itself has come to be conceived as the ultimate fact, and properties of matter have been explained in terms of it. Thus energy, mass, matter, chemical elements are now expressed as electromagnetic phenomena. The ether is still there for those who like it, but it merely serves to allay the tortures of a metaphysical craving.

But Einstein and Minkowski have gone farther. Hitherto time and space have been treated as separate and independent factors in the scheme of things. They have combined them. This is a complete refashioning of older ideas and is in many ways much more consonant with the Cartesian point of view.

The world as we observe it involves process and extension. Hitherto process has been identified with serial time, and extension with space. But this neglects the fact that there is an extension of time. Conceive any ultimate concrete fact as an extended process. If you have lost process or lost extension, you know that you are dealing with abstraction. What is going on here in this room is extended process. Extension and process are each abstractions. But these abstractions can be made in different ways. The space which we apprehend as extension without process and the time which we apprehend as serial process without spatial extension are not each unique. In different circumstances we affix different meanings to the notion of space, and different meanings to the correlative notion of time. In respect to space there is no paradox in this assertion. For us the space of this room is a definite volume; for a man in the sun the room is sweeping through space. But it is paradoxical to hold that the serial process which we apprehend as time is different from the serial process which the man in the sun apprehends as time. Yet if you do that, you can introduce mathematical formulæ expressing spatio-temporal measurements which at one sweep explain a whole multitude of perplexing scientific observation. In fact the formulæ practically have to be admitted, and the theory is the simplest explanation of them. Also philosophically the closer association of time and space is a great advantage.

We now come back to Descartes. He conceived extension as essentially a quality of matter. Generalize his idea: the ultimate fact is not static matter but the flux of physical existence: call any part of this flux, with all its fullness of content and happening, an event: extension is essentially a quality of events and so is process. But the becomingness of nature is not to be constricted within one serial linear procession of time. It requires an indefinite number of such processions to express the complete vision.

If this line of thought, which is that underlying the modern relativity, be admitted, the whole synthesis of the seventeenth century has to be recast. Its Time, its Space, and its Matter are in the melting-pot—and there we must leave them.

Axioms of Geometry

THEORIES OF SPACE

UNTIL THE DISCOVERY of the non-Euclidean geometries (Lobatchewsky, 1826 and 1829; J. Bolyai, 1832; B. Riemann, 1854), geometry was universally considered as being exclusively the science of existent space. (See section VI *Non-Euclidean Geometry*.) In respect to the science, as thus conceived, two controversies may be noticed. First, there is the controversy respecting the absolute and relational theories of space. According to the absolute theory, which is the traditional view (held explicitly by Newton), space has an existence, in some sense whatever it may be, independent of the bodies which it contains. The bodies occupy space, and it is not intrinsically unmeaning to say that any definite body occupies *this* part of space, and not *that* part of space, without reference to other bodies occupying space. According to the relational theory of space, of which the chief exponent was Leibniz,[1] space is nothing but a certain assemblage of the relations between the various particular bodies in space. The idea of space with no bodies in it is absurd. Accordingly there can be no meaning in saying that a body is *here* and not *there*, apart from a reference to the other bodies in the universe. Thus, on this theory, absolute motion is intrinsically unmeaning. It is admitted on all hands that in practice only relative motion is directly measurable.

[1] For an analysis of Leibniz's ideas on space, cf. B. Russell, *The Philosophy of Leibniz*, chs. viii-x.

Newton, however, maintains in the *Principia* (scholium to the 8th definition) that it is indirectly measurable by means of the effects of "centrifugal force" as it occurs in the phenomena of rotation. This irrelevance of absolute motion (if there be such a thing) to science has led to the general adoption of the relational theory by modern men of science. But no decisive argument for either view has at present been elaborated.[2] Kant's view of space as being a form of perception at first sight appears to cut across this controversy. But he, saturated as he was with the spirit of the Newtonian physics, must (at least in both editions of the *Critique*) be classed with the upholders of the absolute theory. The form of perception has a type of existence proper to itself independently of the particular bodies which it contains. For example, he writes:[3]

> "Space does not represent any quality of objects by themselves, or objects in their relation to one another, i.e., space does not represent any determination which is inherent in the objects themselves, and would remain, even if all subjective conditions of intuition were removed."

Axioms

The second controversy is that between the view that the axioms applicable to space are known only from experience, and the view that in some sense these axioms are given *a priori*. Both these views, thus broadly stated, are capable of various subtle modifications, and a discussion of them would merge into a general treatise on epistemology. The cruder forms of the *a priori* view have been made quite untenable by the modern mathematical discoveries. Geometers now profess ignorance in many respects of the exact axioms which apply to

[2] Cf. Hon. Bertrand Russell, "Is Position in Time and Space Absolute or Relative?" *Mind*, n.s. vol. 10 (1901), and A. N. Whitehead, "Mathematical Concepts of the Material World," *Phil. Trans.* (1906), p. 205.

[3] Cf. *Critique of Pure Reason*, 1st section; "Of Space," conclusion A, Max Müller's translation.

existent space, and it seems unlikely that a profound study of the question should thus obliterate *a priori* intuitions.

Another question irrelevant to this article, but with some relevance to the above controversy, is that of the derivation of our perception of existent space from our various types of sensation. This is a question for psychology.[4]

Definition of Abstract Geometry.—Existent space is the subject matter of only one of the applications of the modern science of abstract geometry, viewed as a branch of pure mathematics. Geometry has been defined [5] as "the study of series of two or more dimensions." It has also been defined [6] as "the science of cross classification." These definitions are founded upon the actual practice of mathematicians in respect to their use of the term "Geometry." Either of them brings out the fact that geometry is not a science with a determinate subject matter. It is concerned with any subject matter to which the formal axioms may apply. Geometry is not peculiar in this respect. All branches of pure mathematics deal merely with types of relations. Thus the fundamental ideas of geometry (e.g., those of *points* and of *straight lines*) are not ideas of determinate entities, but of any entities for which the axioms are true. And a set of formal geometrical axioms cannot in themselves be true or false, since they are not determinate propositions, in that they do not refer to a determinate subject matter. The axioms are propositional functions.[7] When a set of axioms is given, we can ask (1) whether they are consistent, (2) whether their "existence theorem" is proved, (3) whether they are independent. Axioms are consistent when the contradictory of any axiom cannot

---

[4] Cf. Ernst Mach, *Erkenntniss und Irrtum* (Leipzig); the relevant chapters are translated by T. J. McCormack, *Space and Geometry* (London, 1906); also A. Meinong, *Über die Stellung der Gegenstandstheorie im System der Wissenschaften* (Leipzig, 1907).

[5] Cf. Russell, *Principles of Mathematics*, § 352 (Cambridge, 1903).

[6] Cf. A. N. Whitehead, *The Axioms of Projective Geometry*, § 3 (Cambridge, 1906).

[7] Cf. Russell, *Princ. of Math.*, ch. i.

be deduced from the remaining axioms. Their existence theorem is the proof that they are true when the fundamental ideas are considered as denoting some determinate subject matter, so that the axioms are developed into determinate propositions. It follows from the logical law of contradiction that the proof of the existence theorem proves also the consistency of the axioms. This is the only method of proof of consistency. The axioms of a set are independent of each other when no axiom can be deduced from the remaining axioms of the set. The independence of a given axiom is proved by establishing the consistency of the remaining axioms of the set, together with the contradictory of the given axiom. The enumeration of the axioms is simply the enumeration of the hypotheses[8] (with respect to the undetermined subject matter) of which some at least occur in each of the subsequent propositions.

Any science is called a "geometry" if it investigates the theory of the classification of a set of entities (the points) into classes (the straight lines), such that (1) there is one and only one class which contains any given pair of entities, and (2) every such class contains more than two members. In the two geometries, important from their relevance to existent space, axioms which secure an order of the points on any line also occur. These geometries will be called "Projective Geometry" and "Descriptive Geometry." In projective geometry any two straight lines in a plane intersect, and the straight lines are closed series which return into themselves, like the circumference of a circle. In descriptive geometry two straight lines in a plane do not necessarily intersect, and a straight line is an open series without beginning or end. Ordinary Euclidean geometry is a descriptive geometry; it becomes a projective geometry when the so-called "points at infinity" are added.

Projective Geometry

Projective geometry may be developed from two undefined fundamental ideas, namely, that of a "point"

[8] Cf. Russell, loc. cit., and G. Frege, "Über die Grundlagen der Geometrie," Jahresber. der Deutsch. Math. Ver. (1906).

and that of a "straight line." These undetermined ideas take different specific meanings for the various specific subject matters to which projective geometry can be applied. The number of the axioms is always to some extent arbitrary, being dependent upon the verbal forms of statement which are adopted. They will be presented [9] here as twelve in number, eight being "axioms of classification," and four being "axioms of order."

Axioms of Classification.—The eight axioms of classification are as follows:

1. Points form a class of entities with at least two members.

2. Any straight line is a class of points containing at least three members.

3. Any two distinct points lie in one and only one straight line.

4. There is at least one straight line which does not contain all the points.

5. If A, B, C are non-collinear points, and A′ is on the straight line BC, and B′ is on the straight line CA, then the straight line AA′ and BB′ possess a point in common.

Definition.—If A, B, C are any three non-collinear points, the *plane* ABC is the class of points lying on the straight lines joining A with the various points on the straight line BC.

6. There is at least one plane which does not contain all the points.

7. There exists a plane α, and a point A not incident in α, such that any point lies in some straight line which contains both A and a point in α.

Definition.—Harm. (ABCD) symbolizes the following conjoint statements: (1) that the points A, B, C, D are collinear, and (2) that a quadrilateral can be found with

[9] This formulation—though not in respect to number—is in all essentials that of M. Pieri, cf. "I principii della Geometria di Posizione," *Accad. R. di Torino* (1898); also cf. Whitehead, *loc. cit.*

one pair of opposite sides intersecting at A, with the other pair intersecting at C, and with its diagonals passing through B and D respectively. Then B and D are said to be "harmonic conjugates" with respect to A and C.

8. Harm. (ABCD) implies that B and D are distinct points.

In the above axioms 4 secures at least two dimensions, axiom 5 is the fundamental axiom of the plane, axiom 6 secures at least three dimensions, and axiom 7 secures at most three dimensions. From axioms 1-5 it can be proved that any two distinct points in a straight line determine that line, that any three non-collinear points in a plane determine that plane, that the straight line containing any two points in a plane lies wholly in that plane, and that any two straight lines in a plane intersect. From axioms 1-6 Desargues's well-known theorem on triangles in perspective can be proved.

The enunciation of this theorem is as follows: if ABC and A'B'C' are two coplanar triangles such that the lines AA', BB', CC' are concurrent, then the three points of intersection of BC and B'C' of CA and C'A', and of AB and A'B' are collinear; and conversely if the three points of intersection are collinear, the three lines are concurrent. The proof which can be applied is the usual projective proof by which a third triangle A''B''C'' is constructed not coplanar with the other two, but in perspective with each of them.

It has been proved[10] that Desargues's theorem cannot be deduced from axioms 1-5, that is, if the geometry be confined to two dimensions. All the proofs proceed by the method of producing a specification of "points" and "straight lines" which satisfies axioms 1-5, and such that Desargues's theorem does not hold.

[10] Cf. G. Peano, "Sui fondamenti della Geometria," p. 73, *Rivista di matematica,* vol. iv (1894), and D. Hilbert, *Grundlagen der Geometrie* (Leipzig, 1899); and R. F. Moulton, "A Simple non-Desarguesian Plane Geometry," *Trans. Amer. Math. Soc.,* vol. iii (1902).

It follows from axioms 1-5 that Harm. (ABCD) implies Harm. (ADCB) and Harm. (CBAD), and that, if A, B, C be any three distinct collinear points, there exists at least one point D such that Harm. (ABCD). But it requires Desargues's theorem, and hence axiom 6, to prove that Harm. (ABCD) and Harm. (ABCD') imply the identity of D and D'.

The necessity for axiom 8 has been proved by G. Fano,[11] who has produced a three dimensional geometry of fifteen points, i.e., a method of cross classification of fifteen entities, in which each straight line contains three points, and each plane contains seven straight lines. In this geometry axiom 8 does not hold. Also from axioms 1-6 and 8 it follows that Harm. (ABCD) implies Harm. (BCDA).

Definitions.—When two plane figures can be derived from one another by a single projection, they are said to be in *perspective*. When two plane figures can be derived one from the other by a finite series of perspective relations between intermediate figures they are said to be *projectively* related. Any property of a plane figure which necessarily also belongs to any projectively related figure, is called a *projective* property.

The following theorem, known from its importance as "the fundamental theorem of projective geometry," cannot be proved [12] from axioms 1-8. The enunciation is: "A projective correspondence between the points on two straight lines is completely determined when the correspondents of three distinct points on one line are determined on the other." This theorem is equivalent[13] (assuming axioms 1-8) to another theorem, known as

---

[11] Cf. "Sui postulati fondamentali della geometria projettiva," *Giorn. di matematica,* vol. xxx (1891); also of Pieri, *loc. cit.,* and Whitehead, *loc. cit.*

[12] Cf. Hilbert, *loc. cit.;* for a fuller exposition of Hilbert's proof cf. K. T. Vahlen, *Abstrakte Geometrie* (Leipzig, 1905), also Whitehead, *loc. cit.*

[13] Cf. H. Wiener, *Jahresber. der Deutsch. Math. Ver.* vol. i (1890); and F. Schur, "Uber den Fundamentalsatz der projectiven Geometrie," *Math. Ann.* vol. li (1899).

Pappus's Theorem, namely: "If l and l' are two distinct coplanar lines, and A, B, C are three distinct points on l, and A', B' C' are three distinct points on l', then the three points of intersection AA' and B'C, of A'B and CC', of BB' and C'A, are collinear." This theorem is obviously Pascal's well-known theorem respecting a hexagon inscribed in a conic, for the special case when the conic has degenerated into the two lines l and l'. Another theorem also equivalent (assuming axioms 1-8) to the fundamental theorem is the following:[14] If the three collinear pairs of points, A and A', B and B', C and C', are such that the three pairs of opposite sides of a complete quadrangle pass respectively through them, i.e. one pair through A and A' respectively, and so on, and if also the three sides of the quadrangle which pass through A, B, and C, are concurrent in one of the corners of the quadrangle, then another quadrangle can be found with the same relation to the three pairs of points, except that its three sides which pass through A, B, and C, are not concurrent.

Thus, if we choose to take any one of these three theorems as an axiom, all the theorems of projective geometry which do not require ordinal or metrical ideas for their enunciation can be proved. Also a conic can be defined as the locus of the points found by the usual construction, based on Pascal's theorem, for points on the conic through five given points. But it is unnecessary to assume here any one of the suggested axioms; for the fundamental theorem can be deduced from the axioms of order together with axioms 1-8.

Axioms of Order.—It is possible to define (cf. Pieri, *loc. cit.*) the property upon which the order of points on a straight line depends. But to secure that this property does in fact range the points in a serial order, some axioms are required. A straight line is to be a closed series; thus, when the points are in order, it requires two points on the line to divide it into two distinct complementary segments, which do not overlap, and

[14] Cf. Hilbert, *loc. cit.*, and Whitehead, *loc. cit.*

together form the whole line. Accordingly the problem of the definition of order reduces itself to the definition of these two segments formed by any two points on the line; and the axioms are stated relatively to these segments.

Definitions.—If A, B, C are three collinear points, the points on the *segment ABC* are defined to be those points such as X, for which there exist two points Y and Y' with the property that Harm. (AYCY') and Harm. (BYXY') both hold. The *supplementary segment ABC* is defined to be the rest of the points on the line. This definition is elucidated by noticing that with our ordinary geometrical ideas, if B and X are any two points between A and C, then the two pairs of points, A and C, B and X, define an involution with real double points, namely, the Y and Y' of the above definition. The property of belonging to a segment ABC is projective, since the harmonic relation is projective.

The first three axioms of order (cf. Pieri, *loc. cit.*) are:

9. If A, B, C are three distinct collinear points, the supplementary segment ABC is contained within the segment BCA.

10. If A, B, C are three distinct collinear points, the common part of the segments BCA and CAB is contained in the supplementary segment ABC.

11. If A, B, C are three distinct collinear points, and D lies in the segment ABC, then the segment ADC is contained within the segment ABC.

From these axioms all the usual properties of a closed order follow. It will be noticed that, if A, B, C are any three collinear points, C is necessarily traversed in passing from A to B by one route along the line, and is not traversed in passing from A to B along the other route. Thus there is no meaning, as referred to closed straight lines, in the simple statement that C lies between A and B. But there may be a relation of separation between two pairs of collinear points, such as A and C, and B and D. The couple B and D is said to separate A and C, if the four points are collinear and D lies in the segment complementary to the segment ABC. The property of the separation of pairs of points by pairs of points is projective. Also it can be proved that Harm. (ABCD) implies that B and D separate A and C.

Definitions.—A series of entities arranged in a serial order, open or closed, is said to be *compact,* if the series contains no immediately consecutive entities, so that in traversing the series from any one entity to any other entity it is necessary to pass through entities distinct from either. It was the merit of R. Dedekind and of G. Cantor explicitly to formulate another fundamental property of series. The Dedekind property[15] as applied to an open series can be defined thus: An open series possesses the Dedekind property, if, however, it be divided into two mutually exclusive classes u and v, which (1) contain between them the whole series, and (2) are such that every member of u precedes in the serial order every member of v, there is always a member of the series, belonging to one of the two, u or v, which precedes every member of v (other than itself if it belong to v), and also succeeds every member of u (other than itself if it belong to u). Accordingly in an open series with the Dedekind property there is always a member of the series marking the junction of two classes such as u and v. An open series is *continuous* if it is compact and possesses the Dedekind property. A closed series can always be transformed into an open series by taking any arbitrary member as the first term and by taking one of the two ways round as the ascending order of the series. Thus the definitions of compactness and of the Dedekind property can be at once transferred to a closed series.

12. The last axiom of order is that there exists at least one straight line for which the point order possesses the Dedekind property.

[15] Cf. Dedekind, *Stetigkeit und irrationale Zahlen* (1872).

It follows from axioms 1-12 by projection that the Dedekind property is true for all lines. Again the *harmonic system ABC,* where A, B, C are collinear points, is defined [16] thus: take the harmonic conjugates A', B', C' of each point with respect to the other two, again take the harmonic conjugates of each of the six points A, B, C, A', B', C' with respect to each pair of the remaining five, and proceed in this way by an unending series of steps. The set of points thus obtained is called the harmonic system ABC. It can be proved that a harmonic system is compact, and that every segment of the line containing it possesses members of it. Furthermore, it is easy to prove that the fundamental theorem holds for harmonic systems, in the sense that, if A, B, C are three points on a line *l,* and A', B', C' are three points on a line *l',* and if by any two distinct series of projections, A, B, C are projected into A', B', C', then any point of the harmonic system ABC corresponds to the same point of the harmonic system A'B'C' according to both the projective relations which are thus established between *l* and *l'.* It now follows immediately that the fundamental theorem must hold for all the points on the lines *l* and *l',* since (as has been pointed out) harmonic systems are "everywhere dense" on their containing lines. Thus the fundamental theorem follows from the axioms of order.

A system of numerical co-ordinates can now be introduced, possessing the property that linear equations represent planes and straight lines. The outline of the argument by which this remarkable problem (in that "distance" is as yet undefined) is solved, will now be given. It is first proved that the points on any line can in a certain way be definitely associated with all the positive and negative real numbers, so as to form with them a one-one correspondence. The arbitrary elements in the establishment of this relation are the points on the line associated with 0, 1 and ∞.

---

[16] Cf. v. Staudt, *Geometrie der Lage* (1847).

This association[17] is most easily effected by considering a class of projective relations of the line with itself, called by F. Schur (*loc. cit.*) *prospectivities*.

Let *l* be the given line, *m* and *n* any two lines intersecting at U on *l*, S and S′ two points on *n*. Then a projective relation between *l* and itself is formed by projecting *l* from S on to *m*, and then by projecting *m* from S′ back on to *l*. All such projective relations, however *m*, *n*, S and S′ be varied, are called "prospectivities," and U is the double point of the prospectivity. If a point O on *l* is related to A by a prospectivity, then all prospectivities, which (1) have the same double point U, and (2) relate O to A, give the same correspondent (Q, in figure) to any point P on the line *l*; in fact they are all the same prospectivity, however *m*, *n*, S, and S′ may have been varied subject to these conditions. Such a prospectivity will be denoted by (OAU^2).

The sum of two prospectivities, written $(OAU^2) + (OBU^2)$, is defined to be that transformation of the line *l* into itself which is obtained by first applying the prospectivity (OAU^2) and then applying the prospectivity (OBU^2). Such a transformation, when the two summands have the same double point, is itself a prospectivity with that double point.

With this definition of addition it can be proved that prospectivities with the same double point satisfy all the axioms of magnitude. Accordingly they can be associated in a one-one correspondence with the positive and negative real numbers. Let E be any point on *l*, distinct from O and U. Then the prospectivity (OEU^2) is associated with unity, the prospectivity (OOU^2) is associated with zero, and (OUU^2) with ∞. The prospectivities of the type (OPU^2), where P is any point on the segment OEU, correspond to the positive numbers; also

[17] Cf. Pasch, *Vorlesungen über neuere Geometrie* (Leipzig, 1882), a classic work; also Fiedler, *Die darstellende Geometrie* (1st ed., 1871, 3rd ed., 1888); Clebsch, *Vorlesungen über Geometrie*, vol. iii; Hilbert, *loc. cit.*; F. Schur, *Math. Ann.* Bd. lv (1902); Vahlen, *loc. cit.*; Whitehead, *loc. cit.*

if P' is the harmonic conjugate of P with respect to O and U, the prospectivity $(OP'U^2)$ is associated with the corresponding negative number. Then any point P on l is associated with the same number as is the prospectivity (OPU^2).

It can be proved that the order of the numbers in algebraic order of magnitude agrees with the order on the line of the associated points. Let the numbers, assigned according to the preceding specification, be said to be associated with the points according to the "numeration-system (OEU)." The introduction of a co-ordinate system for a plane is now managed as follows: Take any triangle OUV in the plane, and on the lines OU and OV establish the numeration systems (OE_1U) and (OE_2V), where E_1 and E_2 are arbitrarily chosen. Then if M and N are associated with the numbers x and y according to these systems, the co-ordinates of P are x and y. It then follows that the equation of a straight line is of the form $ax + by + c = O$. Both co-ordinates of any point on the line UV are infinite. This can be avoided by introducing homogeneous co-ordinates X, Y, Z, where $x = X/Z$, and $y = Y/Z$, and $Z = O$ is the equation of UV.

The procedure for three dimensions is similar. Let OUVW be any tetrahedron, and associate points on OU, OV, OW with numbers according to the numeration systems (OE_1U), (OE_2V), and (OE_3W). Let the planes VWP, WUP, UVP cut OU, OV, OW in L, M, N respectively; and let x, y, z be the numbers associated with L, M, N respectively. Then P is the point (x, y, z). Also homogeneous co-ordinates can be introduced as before, thus avoiding the infinities on the plane UVW.

The cross ratio of a range of four collinear points can now be defined as a number characteristic of that range. Let the co-ordinates of any point P_r of the range $P_1\ P_2\ P_3\ P_4$ be

$$\frac{\lambda_r a + \mu_r + a'}{\lambda_r + \mu_r},\ \frac{\lambda_r b + \mu_r b'}{\lambda_r + \mu_r},\ \frac{\lambda_r c + \mu_r c'}{\lambda_r + \mu_r},\ (r = 1,\ 2,\ 3,\ 4)$$

and let $(\lambda_r\mu_s)$ be written for $\lambda_r\mu_s-\lambda_s\mu_r$. Then the cross ratio $\{P_1\ P_2\ P_3\ P_4\}$ is defined to be the number $(\lambda_1\mu_2)$ $(\lambda_3\mu_4)\,/\,(\lambda_1\mu_4)\ (\lambda_3\mu_2)$. The equality of the cross ratios of the ranges $(P_1\ P_2\ P_3\ P_4)$ and $(Q_1\ Q_2\ Q_3\ Q_4)$ is proved to be the necessary and sufficient condition for their mutual projectivity. The cross ratios of all harmonic ranges are then easily seen to be all equal to -1, by comparing with the range $(OE_1UE'_1)$ on the axis of x.

Thus all the ordinary propositions of geometry in which distance and angular measure do not enter otherwise than in cross ratios can now be enunciated and proved. Accordingly the greater part of the analytical theory of conics and quadrics belongs to geometry at this stage. The theory of distance will be considered after the principles of descriptive geometry have been developed.

Descriptive Geometry

Descriptive geometry is essentially the science of multiple order for open series. The first satisfactory system of axioms was given by M. Pasch.[18] An improved version is due to G. Peano.[19] Both these authors treat the idea of the class of points constituting the segment lying *between* two points as an undefined fundamental idea. Thus in fact there are in this system two fundamental ideas, namely, of points and of segments. It is then easy enough to define the prolongations of the segments, so as to form the complete straight lines. D. Hilbert's[20] formulation of the axioms is in this respect practically based on the same fundamental ideas. His work is justly famous for some of the mathematical investigations contained in it, but his exposition of the axioms is distinctly inferior to that of Peano. Descriptive geometry can also be considered [21] as the science of a class of relations,

[18] Cf. *loc. cit.*

[19] Cf. *I Principii di geometria* (Turin, 1889) and "Sui fondamenti della geometria," *Rivista di mat.*, vol. iv (1894).

[20] Cf. *loc. cit.*

[21] Cf. Vailati, *Rivista di mat.*, vol. iv, and Russell, *loc. cit.* § 376.

each relation being a two-termed serial relation, as considered in the logic of relations, ranging the points between which it holds into a linear open order. Thus the relations are the straight lines, and the terms between which they hold are the points. But a combination of these two points of view yields[22] the simplest statement of all. Descriptive geometry is then conceived as the investigation of an undefined fundamental relation between three terms (points); and when the relation holds between three points A, B, C, the points are said to be "in the [linear] order ABC."

O. Veblen's axioms and definitions, slightly modified, are as follows:—

1. If the points A, B, C are in the order ABC, they are in the order CBA.

2. If the points A, B, C are in the order ABC, they are not in the order BCA.

3. If the points A, B, C are in the order ABC, A is distinct from C.

4. If A and B are any two distinct points, there exists a point C such that A, B, C are in the order ABC.

Definition.—The *line* AB (A $\pm$ B) consists of A and B, and of all points X in one of the possible orders, ABX, AXB, XAB. The points X in the order AXB constitute the *segment* AB.

5. If points C and D (C $\pm$ D) lie on the line AB, then A lies on the line CD.

6. There exist three distinct points A, B, C not in any of the orders ABC, BCA, CAB.

7. If three distinct points A, B, C do not lie on the same line, and D and E are two distinct points in the orders BCD and CEA, then a point F exists in the order AFB, and such that D, E, F are collinear.

Definition.—If A, B, C are three non-collinear points, the *plane* ABC is the class of points which lie on any one of the lines joining any two of the points belonging

[22] Cf. O. Veblen, "On the Projective Axioms of Geometry," *Trans. Amer. Math. Soc.*, vol. iii (1902).

to the boundary of the triangle ABC, the *boundary* being formed by the segments BC, CA and AB. The *interior* of the triangle ABC is formed by the points in segments such as PQ, where P and Q are points respectively on two of the segments BC, CA, AB.

8. There exists a plane ABC, which does not contain all the points.

Definition.—If A, B, C, D are four non-coplanar points, the space ABCD is the class of points which lie on any of the lines containing two points on the surface of the tetrahedron ABCD, the *surface* being formed by the interiors of the triangles ABC, BCD, DCA, DAB.

9. There exists a space ABCD which contains all the points.

10. The Dedekind property holds for the order of the points on any straight line.

It follows from axioms 1-9 that the points on any straight line are arranged in an open serial order. Also all the ordinary theorems respecting a point dividing a straight line into two parts, a straight line dividing a plane into two parts, and a plane dividing space into two parts, follow.

Again, in any plane α consider a line l and a point A.

Let any point B divide l into two half-lines l_1 and l_2. Then it can be proved that the set of half-lines, emanating from A and intersecting l_1 (such as m), are bounded by two half-lines, of which ABC is one. Let r be the other. Then it can be proved that r does not intersect l_1. Similarly for the half-line, such as n, intersecting l_2. Let s be its bounding half-line. Then two cases are possible. (1) The half-lines r and s are collinear, and together form one complete line. In this case, there is one and only one line (viz., $r + s$) through A and lying in α which does not intersect l. This is the Euclidean case, and the assumption that this case holds is the *Euclidean parallel axiom*. But (2) the half-lines r and s may not be collinear. In this case there will be an infinite number of lines, such as k for instance, containing A and lying in α, which do not intersect l. Then the lines through A in α are divided into two classes by reference to l, namely the *secant* lines which intersect l, and the *non-secant* lines which do not intersect l. The two boundary non-secant lines, of which r and s are respectively halves, may be called the two parallels to l through A.

The perception of the possibility of case 2 constituted the starting-point from which Lobatchewsky constructed the first explicit coherent theory of non-Euclidean geometry, and thus created a revolution in the philosophy of the subject. For many centuries the speculations of mathematicians on the foundations of geometry were almost confined to hopeless attempts to prove the "parallel axiom" without the introduction of some equivalent axiom.[23]

Associated Projective and Descriptive Spaces.—A region of a projective space, such that one, and only one, of the two supplementary segments between any pair of points within it lies entirely within it, satisfies the above axioms (1-10) of descriptive geometry, where the points of the region are the descriptive points, and the portions of straight lines within the region are the descriptive lines. If the excluded part of the original projective space is a single plane, the Euclidean parallel axiom also holds, otherwise it does not hold for the descriptive space of the limited region. Again, conversely, starting from an original descriptive space an associated projective space can be constructed by means of the concept of *ideal points*.[24] These are also called *projective points*, where it is understood that the simple points are the points of the original descriptive space. An *ideal point* is the class of straight lines which is composed of two coplanar lines a and b, together with the lines of inter-

[23] Cf. P. Stäckel and F. Engel, *Die Theorie der Parallellinien von Euklid bis auf Gauss* (Leipzig, 1895).

[24] Cf. Pasch, *loc. cit.,* and R. Bonola, "Sulla introduzione degli enti improprii in geometria projettive," *Giorn. di mat.* vol. xxxviii (1900); and Whitehead, *Axioms of Descriptive Geometry* (Cambridge, 1907).

section of all pairs of intersecting planes which respectively contain a and b, together with the lines of intersection with the plane ab of all planes containing any one of the lines (other than a or b) already specified as belonging to the ideal point. It is evident that, if the two original lines a and b intersect, the corresponding ideal point is nothing else than the whole class of lines which are concurrent at the point ab. But the essence of the definition is that an ideal point has an existence when the lines a and b do not intersect, so long as they are coplanar. An ideal point is termed *proper,* if the lines composing it intersect; otherwise it is *improper.*

A theorem essential to the whole theory is the following: if any two of the three lines a, b, c are coplanar, but the three lines are not all coplanar, and similarly for the lines a, b, d, then c and d are coplanar. It follows that any two lines belonging to an ideal point can be used as the pair of guiding lines in the definition. An ideal point is said to be *coherent* with a plane, if any of the lines composing it lie in the plane. An *ideal line* is the class of ideal points each of which is coherent with two given planes. If the planes intersect, the ideal line is termed *proper,* otherwise it is *improper.* It can be proved that any two planes, with which any two of the ideal points are both coherent, will serve as the guiding planes used in the definition. The ideal planes are defined as in projective geometry, and all the other definitions (for segments, order, etcetera) of projective geometry are applied to the ideal elements. If an ideal plane contains some proper ideal points, it is called *proper,* otherwise it is *improper.* Every ideal plane contains some improper ideal points.

It can now be proved that all the axioms of projective geometry hold of the ideal elements as thus obtained; and also that the order of the ideal points as obtained by the projective method agrees with the order of the proper ideal points as obtained from that of the associated points of the descriptive geometry. Thus a projective space has been constructed out of the ideal elements, and the proper ideal elements correspond element by element with the associated descriptive elements. Thus the proper ideal elements form a region in the projective space within which the descriptive axioms hold. Accordingly, by substituting ideal elements, a descriptive space can always be considered as a region within a projective space. This is the justification for the ordinary use of the "points at infinity" in the ordinary Euclidean geometry; the reasoning has been transferred from the original descriptive space to the associated projective space of ideal elements; and with the Euclidean parallel axiom the improper ideal elements reduce to the ideal points on a single improper ideal plane, namely, the plane at infinity.[25]

Congruence and Measurement.—The property of physical space which is expressed by the term "measurability" has now to be considered. This property has often been considered as essential to the very idea of space. For example, Kant writes,[26] "Space is represented as an infinite given *quantity*." This quantitative aspect of space arises from the measurability of distances, of angles, of surfaces and of volumes. These four types of quantity depend upon the two first among them as fundamental. The measurability of space is essentially connected with the idea of *congruence*, of which the simplest examples are to be found in the proofs of equality by the method of superposition, as used in elementary plane geometry. The mere concepts of "part" and of "whole" must of necessity be inadequate as the foundation of measurement, since we require the comparison as to quantity of regions of space which have no portions in common. The idea of congruence, as exemplified by the method of superposition in geometrical reasoning, appears to be founded upon that of the "rigid body," which moves from one position to another with its internal spatial relations unchanged. But unless there is a previous con-

---

[25] The original idea (confined to this particular case) of ideal points is due to von Staudt (*loc. cit.*).

[26] Cf. *Critique*, "Trans. Aesth." Sect. 1.

cept of the metrical relations between the parts of the body, there can be no basis from which to deduce that they are unchanged.

It would therefore appear as if the idea of the congruence, or metrical equality, of two portions of space (as empirically suggested by the motion of rigid bodies) must be considered as a fundamental idea incapable of definition in terms of those geometrical concepts which have already been enumerated. This was in effect the point of view of Pasch.[27] It has, however, been proved by Sophus Lie[28] that congruence is capable of definition without recourse to a new fundamental idea. This he does by means of his theory of finite continuous groups (see GROUPS, THEORY OF), of which the definition is possible in terms of our established geometrical ideas, remembering that co-ordinates have already been introduced. The displacement of a rigid body is simply a mode of defining to the senses a one-one transformation of all space into itself. For at any point of space a particle may be conceived to be placed, and to be rigidly connected with the rigid body; and thus there is a definite correspondence of any point of space with the new point occupied by the associated particle after displacement. Again two successive displacements of a rigid body from position A to position B, and from position B to position C, are the same in effect as one displacement from A to C. But this is the characteristic "group" property. Thus the transformations of space into itself defined by displacements of rigid bodies form a group.

Call this group of transformations a congruence-group. Now according to Lie a congruence-group is defined by the following characteristics:—

1. A congruence-group is a finite continuous group of one-one transformations, containing the identical transformation.

2. It is a sub-group of the general projective group, i.e. of the group of which any transformation converts

[27] Cf. *loc. cit.*

[28] Cf. *Über die Grundlagen der Geometrie* (Leipzig, *Ber.,* 1890); and *Theorie der Transformationsgruppen* (Leipzig, 1893), vol. iii.

planes into planes, and straight lines into straight lines.

3. An infinitesimal transformation can always be found satisfying the condition that, at least throughout a certain enclosed region, any definite line and any definite point on the line are latent, i.e., correspond to themselves.

4. No infinitesimal transformation of the group exists, such that, at least in the region for which (3) holds, a straight line, a point on it, and a plane through it, shall be latent.

The property enunciated by conditions (3) and (4), taken together, is named by Lie "Free mobility in the infinitesimal." Lie proves the following theorems for a projective space:—

1. If the above four conditions are only satisfied by a group throughout part of projective space, this part either (α) must be the region enclosed by a real closed quadric, or (β) must be the whole of the projective space with the exception of a single plane. In case (α) the corresponding congruence group is the continuous group for which the enclosing quadric is latent; and in case (β) an imaginary conic (with a real equation) lying in the latent plane is also latent, and the congruence group is the continuous group for which the plane and conic are latent.

2. If the above four conditions are satisfied by a group throughout the whole of projective space, the congruence group is the continuous group for which some imaginary quadric (with a real equation) is latent.

By a proper choice of non-homogeneous co-ordinates the equation of any quadrics of the types considered, either in theorem 1 (α), or in theorem 2, can be written in the form $1 + c(x^2 + y^2 + z^2) = 0$, where c is negative for a real closed quadric, and positive for an imaginary quadric. Then the general infinitesimal transformation is defined by the three equations:

$$\left.\begin{aligned} dx/dt &= u - \omega_3 y + \omega_2 z + cx(ux + vy + wz), \\ dy/dt &= v - \omega_1 z + \omega_3 x + cy(ux + vy + wz), \\ dz/dt &= w - \omega_2 x + \omega_1 y + cz(ux + vy + wz). \end{aligned}\right\} \quad \text{(A)}$$

In the case considered in theorem 1 (β), with the proper choice of co-ordinates the three equations defining the general infinitesimal transformations are:

$$dx/dt = u - \omega_3 y + \omega_2 z,$$
$$dy/dt = v - \omega_1 z + \omega_3 x,$$ (B)
$$dz/dt = w - \omega_2 x + \omega_1 y.$$

In this case the latent plane is the plane for which at least one of x, y, z are infinite, that is, the plane $0.x + 0.y + 0.z + a = 0$; and the latent conic is the conic in which the cone $x^2 + y^2 + z^2 = 0$ intersects the latent plane.

It follows from theorems 1 and 2 that there is not one unique congruence group, but an indefinite number of them. There is one congruence-group corresponding to each closed real quadric, one to each imaginary quadric with a real equation, and one to each imaginary conic in a real plane and with a real equation. The quadric thus associated with each congruence-group is called the *absolute* for that group, and in the degenerate case of 1 (β) the absolute is the latent plane together with the latent imaginary conic. If the absolute is real, the congruence-group is *hyperbolic;* if imaginary, it is *elliptic;* if the absolute is a plane and imaginary conic, the group is parabolic. Metrical geometry is simply the theory of the properties of some particular congruence-group selected for study.

The definition of distance is connected with the corresponding congruence-group by two considerations in respect to a range of five points $(A_1, A_2, P_1, P_2, P_3)$, of which A_1 and A_2 are on the absolute.

Let $\{A_1 P_1 A_2 P_2\}$ stand for the cross ratio (as defined above) of the range $(A_1 P_1 A_2 P_2)$, with a similar notation for the other ranges. Then

(1) $\log \{A_1 P_1 A_2 P_2\} + \log \{A_1 P_2 A_2 P_3\} = \log \{A_1 P_1 A_2 P_3\}$, and

(2), if the points A_1, A_2, P_1, P_2 are transformed into A'_1, A'_2, P'_1, P'_2 by any transformation of the congruence-group (α) $\{A_1 P_1 A_2 P_2\} = \{A'_1 P'_1 A'_2 P'_2\}$, since the transformation is projective, and (β) A'_1, A'_2, are on the absolute since A_1 and A_2 are on it. Thus if we define the distance P_1P_2 to be $\frac{1}{2}k \log \{A_1P_1A_2P_2\}$, where A_1 and A_2 are the points in which the line P_1P_2 cuts the absolute, and k is some constant, the two characteristic properties of distance, namely, (1) the addition of consecutive lengths on a straight line, and (2) the invariability of distances during a transformation of the congruence-group, are satisfied. This is the well-known Cayley-Klein projective definition[29] of distance, which was elaborated in view of the addition property alone, previously to Lie's discovery of the theory of congruence-groups. For a hyperbolic group when P_1 and P_2 are in the region enclosed by the absolute, $\log \{A_1P_1A_2P_2\}$ is real, and therefore k must be real. For an elliptic group A_1 and A_2 are conjugate imaginaries, and $\log \{A_1P_1A_2P_2\}$ is a pure imaginary, and k is chosen to be $\varkappa/\iota$, where $\varkappa$ is real and $\iota = \sqrt{-}$.

Similarly the angle between two planes, p_1 and p_2, is defined to be $(1/2\iota) \log (t_1p_1t_2p_2)$, where t_1 and t_2 are tangent planes to the absolute through the line p_1p_2. The planes t_1 and t_2 are imaginary for an elliptic group, and also for an hyperbolic group when the planes p_1 and p_2 intersect at points within the region enclosed by the absolute. The development of the consequences of these metrical definitions is the subject of non-Euclidean geometry.

The definitions for the parabolic case can be arrived at as limits of those obtained in either of the other two cases by making k ultimately to vanish. It is also obvious that, if P_1 and P_2 be the points (x_1, y_1, z_1) and (x_2, y_2, z_2), it follows from equations (B) above that $\{ (x_1 - x_2)^2 + (y_1 - y_2)^2 + (z_1 - z_2)^2 \}^{\frac{1}{2}}$ is unaltered by a congruence transformation and also satisfies the addition property for collinear distances. Also the previous definition of an angle can be adapted to this case, by making t_1 and t_2 to be the tangent planes through the line p_1p_2 to the imaginary conic. Similarly if p_1 and p_2 are inter-

[29] Cf. A. Cayley, "A Sixth Memoir on Quantics," *Trans. Roy. Soc.*, 1859, and *Coll. Papers*, vol. ii; and F. Klein, *Math. Ann.* vol. iv, 1871.

secting lines, the same definition of an angle holds, where t_1 and t_2 are now the lines from the point p_1p_2 to the two points where the plane p_1p_2 cuts the imaginary conic. These points are in fact the "circular points at infinity" on the plane. The development of the consequences of these definitions for the parabolic case gives the ordinary Euclidean metrical geometry.

Thus the only metrical geometry for the whole of projective space is of the elliptic type. But the actual measure-relations (though not their general properties) differ according to the elliptic congruence-group selected for study. In a descriptive space a congruence-group should possess the four characteristics of such a group throughout the whole of the space. Then form the associated ideal projective space. The associated congruence-group for this ideal space must satisfy the four conditions throughout the region of the proper ideal points. Thus the boundary of this region is the absolute. Accordingly there can be no metrical geometry for the whole of a descriptive space unless its boundary (in the associated ideal space) is a closed quadric or a plane. If the boundary is a closed quadric, there is one possible congruence-group of the hyperbolic type. If the boundary is a plane (the plane at infinity), the possible congruence-groups are parabolic; and there is a congruence-group corresponding to each imaginary conic in this plane, together with a Euclidean metrical geometry corresponding to each such group. Owing to these alternative possibilities, it would appear to be more accurate to say that systems of quantities can be found in a space, rather than that space is a quantity.

Lie has also deduced[30] the same results with respect to congruence-groups from another set of defining properties, which explicitly assume the existence of a quantitative relation (the distance) between any two points, which is invariant for any transformation of the congruence-group.[31]

[30] Cf. *loc. cit.*

[31] For similar deductions from a third set of axioms, suggested in essence by Peano, *Riv. mat.*, vol. iv, *loc. cit.*, cf. Whitehead, *Desc. Geom.*, *loc. cit.*

The above results, in respect to congruence and metrical geometry, considered in relation to existent space, have led to the doctrine[32] that it is intrinsically unmeaning to ask which system of metrical geometry is true of the physical world. Any one of these systems can be applied, and in an indefinite number of ways. The only question before us is one of convenience in respect to simplicity of statement of the physical laws. This point of view seems to neglect the consideration that science is to be relevant to the definite perceiving minds of men; and that (neglecting the ambiguity introduced by the invariable slight inexactness of observation which is not relevant to this special doctrine) we have, in fact, presented to our senses a definite set of transformations forming a congruence-group, resulting in a set of measure relations which are in no respect arbitrary. Accordingly our scientific laws are to be stated relevantly to that particular congruence-group. Thus the investigation of the type (elliptic, hyperbolic or parabolic) of this special congruence-group is a perfectly definite problem, to be decided by experiment. The consideration of experiments adapted to this object requires some development of non-Euclidean geometry (see section VI, *Non-Euclidean Geometry*). But if the doctrine means that, assuming some sort of objective reality for the material universe, beings can be imagined, to whom *either* all congruence-groups are equally important, *or* some other congruence-group is specially important, the doctrine appears to be an immediate deduction from the mathematical facts. Assuming a definite congruence-group, the investigation of surfaces (or three-dimensional loci in space of four dimensions) with geodesic geometries of the form of metrical geometries of other types of congruence-groups forms an important chapter of non-Euclidean geometry. Arising from this investigation there is a widely-spread fallacy, which has found its way into many philosophic writings, namely, that the possibility of the geometry of existent three-dimensional space being other than Euclidean depends on the physical

[32] Cf. H. Poincaré, *La Science et l'hypothèse*, ch. iii.

existence of Euclidean space of four or more dimensions. The foregoing exposition shows the baselessness of this idea.

BIBLIOGRAPHY.—For an account of the investigations on the axioms of geometry during the Greek period, see M. Cantor, *Vorlesungen über die Geschichte der Mathematik;* Bd. i and iii; T. L. Heath, *The Thirteen Books of Euclid's Elements, a New Translation from the Greek, with Introductory Essays and Commentary, Historical, Critical and Explanatory* (Cambridge, 1908)—this work is the standard source of information; W. B. Frankland, *Euclid, Book I, with a Commentary* (Cambridge, 1905)—the commentary contains copious extracts from the ancient commentators. The next period of really substantive importance is that of the eighteenth century. The leading authors are: G. Saccheri, S.J., *Euclides ab omni naevo vindicatus* (Milan, 1733). Saccheri was an Italian Jesuit who unconsciously discovered non-Euclidean geometry in the course of his efforts to prove its impossibility. J. H. Lambert, *Theorie der Parallellinien* (1766); A. M. Legendre, *Éléments de géométrie* (1794). An adequate account of the above authors is given by P. Stäckel and F. Engel, *Die Theorie der Parallellinien von Euklid bis auf Gauss* (Leipzig, 1895). The next period of time (roughly from 1800 to 1870) contains two streams of thought, both of which are essential to the modern analysis of the subject. The first stream is that which produced the discovery and investigation of non-Euclidean geometries, the second stream is that which has produced the geometry of position, comprising both projective and descriptive geometry not very accurately discriminated. The leading authors on non-Euclidean geometry are K. F. Gauss, in private letters to Schumacher, cf. Stäckel and Engel, *loc. cit.;* N. Lobatchewsky, rector of the university of Kazan, to whom the honour of the effective discovery of non-Euclidean geometry must be assigned. His first publication was at Kazan in 1826. His various memoirs have been re-edited by Engel; cf. *Urkunden zur Geschichte der nichteuklidischen Geometrie* by Stäckel and Engel, vol. i, "Lobatchewsky." J. Bolyai discovered non-Euclidean geometry apparently in independence of Lobatchewsky. His memoir was published in 1831 as an appendix to a work by his father W. Bolyai, *Tentamen juventutem*. . . . This memoir has been separately edited by J. Frischauf, *Absolute Geometrie nach J. Bolyai* (Leipzig, 1872); B. Riemann, *Über die Hypothesen, welche der Geometrie zu Grunde liegen* (1854); cf. *Gesamte Werke,* a translation in *The Collected Papers* of W. K. Clifford. This is a fundamental memoir on the subject and must rank with the work of Lobatchewsky. Riemann discovered elliptic metrical geometry, and Lobatchewsky hyperbolic geometry. A full account of Riemann's ideas, with the subsequent developments due to Clifford, F. Klein and W. Killing, will be found in *The Boston Colloquium for* 1903 (New York, 1905), article "Forms of Non-Euclidean Space," by F. S. Woods. A. Cayley, *loc. cit.* (1859), and F. Klein, "Über die sogenannte nichteuklidische Geometrie," *Math. Annal.* vols. iv and vi (1871 and 1872), between them elaborated the projective theory of distance; H. Helmholtz, "Über die thatsächlichen Grundlagen der Geometrie" (1866), and "Über die Thatsachen, die der Geometrie zu Grunde liegen" (1868), both in his *Wissenschaftliche Abhandlungen,* vol. ii, and S. Lie, *loc. cit.* (1890 and 1893), between them elaborated the group theory of congruence.

The numberless works which have been written to suggest equivalent alternatives to Euclid's parallel axioms may be neglected as being of trivial importance, though many of them are marvels of geometric ingenuity.

The second stream of thought confined itself within the circle of ideas of Euclidean geometry. Its origin was mainly due to a succession of great French mathematicians, for example, G. Monge, *Géométrie descriptive* (1800); J. V. Poncelet, *Traité des propriétés projectives des figures* (1822); M. Chasles, *Aperçu historique sur l'origine et le développement des méthodes en géométrie* (Bruxelles, 1837) and *Traité de géométrie supérieure* (Paris, 1852); and many others. But the works which have been, and are still, of decisive influence on thought as a store-house of ideas relevant to the foundations of geometry are K. G. C. von Staudt's two works, *Geometrie der Lage* (Nürnberg, 1847); and *Beiträge zur Geometrie der Lage* (Nürnberg, 1856, 3rd ed. 1860).

The final period is characterized by the successful production of exact systems of axioms, and by the final solution of problems which have occupied mathematicians for two thousand years. The successful analysis of the ideas involved in serial continuity is due to R. Dedekind, *Stetigkeit und irrationale Zahlen* (1872), and to G. Cantor, *Grundlagen einer allgemeinen Mannigfaltigkeitslehre* (Leipzig, 1883), and *Acta math.* vol. 2.

Complete systems of axioms have been stated by M. Pasch, *loc. cit.;* G. Peano, *loc. cit.;* M. Pieri, *loc. cit.;* B. Russell, *Principles of Mathematics;* O. Veblen, *loc. cit.;* and by G. Veronese in his treatise, *Fondamenti di geometria* (Padua, 1891; German transl. by A. Schepp, *Grundzüge der Geometrie,* Leipzig, 1894). Most of the leading memoirs on special questions involved have been cited in the text; in addition there may be mentioned M. Pieri, "Nuovi principii di geometria projettiva complessa," *Trans. Accad. R. d. Sci.* (Turin, 1905); E. H. Moore, "On the Projective Axioms of Geometry," *Trans. Amer. Math. Soc.,* 1902; O. Veblen and W. H. Bussey, "Finite Projective Geometries," *Trans. Amer. Math. Soc.,* 1905; A. B. Kempe, "On the Relation between the Logical Theory of Classes and the Geometrical Theory of Points," *Proc. Lond. Math. Soc.,* 1890; J. Royce, "The Relation of the Principles of Logic to the Foundations of Geometry," *Trans. of Amer. Math. Soc.,* 1905; A. Schoenflies, "Über die Möglichkeit einer projectiven Geometrie bei transfiniter (nichtarchimedischer) Massbestimmung," *Deutsch. M.-V. Jahresb.,* 1906.

For general expositions of the bearings of the above investigations, cf. Hon. Bertrand Russell, *loc. cit.;* L. Couturat, *Les Principes des mathématiques* (Paris, 1905); H. Poincaré, *loc. cit.;* Russell and Whitehead, *Principia mathematica* (Cambridge Univ. Press). The philosophers whose views on space and geometric truth deserve especial study are Descartes, Leibniz, Hume, Kant and J. S. Mill. (A. N. W.)

Mathematics

MATHEMATICS (Gr. μαξηματική, *sc.* τέχνη or ἐπιστήμη; from μάξημα, "learning" or "science"), the general term for the various applications of mathematical thought, the traditional field of which is number and quantity. It has been usual to define mathematics as "the science of discrete and continuous magnitude." Even Leibniz,[1] who initiated a more modern point of view, follows the tradition in thus confining the scope of mathematics properly so called, while apparently conceiving it as a department of a yet wider science of reasoning. A short consideration of some leading topics of the science will exemplify both the plausibility and inadequacy of the above definition. Arithmetic, algebra, and the infinitesimal calculus are sciences directly concerned with integral numbers, rational (or fractional) numbers, and real numbers generally, which include incommensurable numbers. It would seem that "the general theory of discrete and continuous quantity" is the exact description of the topics of these sciences. Furthermore, can we not complete the circle of the mathematical sciences by adding geometry? Now geometry deals with points, lines, planes and cubic contents. Of these all except points are quantities: lines involve lengths, planes involve areas, and cubic contents involve volumes. Also, as the Cartesian geometry shows, all the relations between points are expressible in terms of geometric quantities. Ac-

[1] Cf. *La Logique de Leibniz*, ch. vii, by L. Couturat (Paris, 1901).

cordingly, at first sight it seems reasonable to define geometry in some such way as "the science of dimensional quantity." Thus every subdivision of mathematical science would appear to deal with quantity, and the definition of mathematics as "the science of quantity" would appear to be justified. We have now to consider the reasons for rejecting this definition as inadequate.

Types of Critical Questions.—What are numbers? We can talk of five apples and ten pears. But what are "five" and "ten" apart from the apples and pears? Also in addition to the cardinal numbers there are the ordinal numbers: the fifth apple and the tenth pear claim thought. What is the relation of "the fifth" and "the tenth" to "five" and "ten"? "The first rose of summer" and "the last rose of summer" are parallel phrases, yet one explicitly introduces an ordinal number and the other does not. Again, "half a foot" and "half a pound" are easily defined. But in what sense is there "a half," which is the same for "half a foot" as "half a pound"? Furthermore, incommensurable numbers are defined as the limits arrived at as the result of certain procedures with rational numbers. But how do we know that there is anything to reach? We must know that $\sqrt{2}$ exists before we can prove that any procedure will reach it. An expedition to the North Pole has nothing to reach unless the earth rotates.

Also in geometry, what is a point? The straightness of a straight line and the planeness of a plane require consideration. Furthermore, "congruence" is a difficulty. For when a triangle "moves," the points do not move with it. So what is it that keeps unaltered in the moving triangle? Thus the whole method of measurement in geometry as described in the elementary textbooks and the older treatises is obscure to the last degree. Lastly, what are "dimensions"? All these topics require thorough discussion before we can rest content with the definition of mathematics as the general science of magnitude; and by the time they are discussed the definition has evaporated. An outline of the modern answers to questions such as the above will now be given. A critical defence of them would require a volume.[2]

Cardinal Numbers.—A one-one relation between the members of two classes α and β is any method of correlating all the members of α to all the members of β, so that any member of α has one and only one correlate in β, and any member of β has one and only one correlate in α. Two classes between which a one-one relation exists have the same cardinal number and are called cardinally similar; and the cardinal number of the class α is a certain class whose members are themselves classes —namely, it is the class composed of all those classes for which a one-one correlation with α exists. Thus the cardinal number of α is itself a class, and furthermore α is a member of it. For a one-one relation can be established between the members of α and α by the simple process of correlating each member of α with itself. Thus the cardinal number one is the class of unit classes, the cardinal number two is the class of doublets, and so on. Also a unit class is any class with the property that it possesses a member x such that, if y is any member of the class, then x and y are identical. A doublet is any class which possesses a member x such that the modified class formed by all the other members except x is a unit class. And so on for all the finite cardinals, which are thus defined successively. The cardinal number zero is the class of classes with no members; but there is only one such class, namely—the null class. Thus this cardinal number has only one member. The operations of addition and multiplication of two given cardinal numbers can be defined by taking two classes α and β, satisfying the conditions (1) that their cardinal numbers are respectively the given numbers, and (2) that they contain no member in common, and then by defining by reference to α and β two other suitable classes whose cardinal numbers are defined to be respectively the required sum and product of the cardinal numbers in question. We need not here consider the details of this process.

[2] Cf. *The Principles of Mathematics*, by Bertrand Russell (Cambridge, 1903).

With these definitions it is now possible to *prove* the following six premises applying to finite cardinal numbers, from which Peano[3] has shown that all arithmetic can be deduced:—

i. Cardinal numbers form a class.

ii. Zero is a cardinal number.

iii. If a is a cardinal number, $a + 1$ is a cardinal number.

iv. If s is any class and zero is a member of it, also if when x is a cardinal number and a member of s, also $x + 1$ is a member of s, then the whole class of cardinal numbers is contained in s.

v. If a and b are cardinal numbers, and $a + 1 = b + 1$, then $a = b$.

vi. If a is a cardinal number, then $a + 1 \neq 0$.

It may be noticed that (iv) is the familiar principle of mathematical induction. Peano in an historical note refers its first explicit employment, although without a general enunciation, to Maurolycus in his work, *Arithmeticorum libri duo* (Venice, 1575).

But now the difficulty of confining mathematics to being the science of number and quantity is immediately apparent. For there is no self-contained science of cardinal numbers. The proof of the six premises requires an elaborate investigation into the general properties of classes and relations which can be deduced by the strictest reasoning from our ultimate logical principles. Also it is purely arbitrary to erect the consequences of these six principles into a separate science. They are excellent principles of the highest value, but they are in no sense the necessary premises which must be proved before any other propositions of cardinal numbers can be established. On the contrary, the premises of arithmetic can be put in other forms, and, furthermore, an indefinite number of propositions of arithmetic can be proved directly from logical principles without men-

[3] Cf. *Formulaire mathématique* (Turin, ed. of 1903); earlier formulations of the bases of arithmetic are given by him in the editions of 1898 and of 1901. The variations are only trivial.

tioning them. Thus, while arithmetic may be defined as that branch of deductive reasoning concerning classes and relations which is concerned with the establishment of propositions concerning cardinal numbers, it must be added that the introduction of cardinal numbers makes no great break in this general science. It is no more than an interesting subdivision in a general theory.

Ordinal Numbers.—We must first understand what is meant by "order," that is, by "serial arrangement." An order of a set of things is to be sought in that relation holding between members of the set which constitutes that order. The set viewed as a class has many orders. Thus the telegraph posts along a certain road have a space-order very obvious to our senses; but they have also a time-order according to dates of erection, perhaps more important to the postal authorities who replace them after fixed intervals. A set of cardinal numbers have an order of magnitude, often called *the* order of the set because of its insistent obviousness to us; but, if they are the numbers drawn in a lottery, their time-order of occurrence in that drawing also ranges them in an order of some importance. Thus the order is defined by the "serial" relation. A relation (R) is serial[4] when (1) it implies diversity, so that, if x has the relation R to y, x is diverse from y; (2) it is transitive, so that if x has the relation R to y, and y to z, then x has the relation R to z; (3) it has the property of connexity, so that if x and y are things to which any things bear the relation R, or which bear the relation R to any things, then *either* x is identical with y, *or* x has the relation R to y, *or* y has the relation R to x. These conditions are necessary and sufficient to secure that our ordinary ideas of "preceding" and "succeeding" hold in respect to the relation R. The "field" of relation R is the class of things ranged in order by it. Two relations R and R' are said to be ordinally similar, if a one-one relation holds between the members of the two fields of R and R', such that if x and y are any two members of the field of R, such that x has the relation R to y, and if x' and y' are

[4] Cf. Russell, *loc. cit.*, pp. 199-256.

the correlates in the field of R' of x and y, then in all such cases x' has the relation R' to y', and conversely, interchanging the dashes on the letters, i.e. R and R', x and x', etcetera. It is evident that the ordinal similarity of two relations implies the cardinal similarity of their fields, but not conversely. Also, two relations need not be serial in order to be ordinally similar; but if one is serial, so is the other. The relation-number of a relation is the class whose members are all those relations which are ordinally similar to it. This class will include the original relation itself. The relation-number of a relation should be compared with the cardinal number of a class. When a relation is serial its relation-number is often called its serial type. The addition and multiplication of two relation-numbers is defined by taking two relations R and S, such that (1) their fields have no terms in common; (2) their relation-numbers are the two relation-numbers in question, and then by defining by reference to R and S two other suitable relations whose relation-numbers are defined to be respectively the sum and product of the relation-numbers in question. We need not consider the details of this process. Now if n be any finite cardinal number, it can be proved that the class of those serial relations, which have a field whose cardinal number is n, is a relation-number. This relation-number is the ordinal number corresponding to n; let it be symbolized by $\dot{n}$. Thus, corresponding to the cardinal numbers 2, 3, 4 . . . there are the ordinal numbers $\dot{2}, \dot{3}, \dot{4}$ . . . The definition of the ordinal number $\dot{1}$ requires some little ingenuity owing to the fact that no serial relation can have a field whose cardinal number is 1; but we must omit here the explanation of the process. The ordinal number 0 is the class whose sole member is the null relation—that is, the relation which never holds between any pair of entities. The definitions of the finite ordinals can be expressed without use of the corresponding cardinals, so there is no essential priority of cardinals to ordinals. Here also it can be seen that the science of the finite ordinals is a particular subdivision of the general theory of classes and relations. Thus the illusory nature of the traditional definition of mathematics is again illustrated.

Cantor's Infinite Numbers.—Owing to the correspondence between the finite cardinals and the finite ordinals, the propositions of cardinal arithmetic and ordinal arithmetic correspond point by point. But the definition of the cardinal number of a class applies when the class is not finite, and it can be proved that there are different infinite cardinal numbers, and that there is a least infinite cardinal, now usually denoted by $\aleph_0$, where $\aleph$ is the Hebrew letter aleph. Similarly, a class of serial relations, called *well-ordered* serial relations, can be defined, such that their corresponding relation-numbers include the ordinary finite ordinals, but also include relation-numbers which have many properties like those of the finite ordinals, though the fields of the relations belonging to them are not finite. These relation-numbers are the infinite ordinal numbers. The arithmetic of the infinite cardinals does not correspond to that of the infinite ordinals. The theory of these extensions of the ideas of number is dealt with in the article. It will suffice to mention here that Peano's fourth premiss of arithmetic does not hold for infinite cardinals or for infinite ordinals. Contrasting the above definitions of number, cardinal and ordinals, with the alternative theory that number is an ultimate idea incapable of definition, we notice that our procedure exacts a greater attention combined with a smaller credulity; for every idea, assumed as ultimate, demands a separate act of faith.

The Data of Analysis.—Rational numbers and real numbers in general can now be defined according to the same general method. If m and n are finite cardinal numbers, the rational number m/n is the relation which any finite cardinal number x bears to any finite cardinal number y when $n \times x = m \times y$. Thus the rational number one, which we will denote by 1_r, is not the cardinal number 1; for 1_r is the relation $1/1$ as defined above, and is thus a relation holding between certain pairs of cardinals. Similarly, the other rational integers must be distinguished from the corresponding cardinals. The arithmetic of rational numbers is now established by means of appropriate definitions, which indicate the entities meant by the operations of addition and multiplication. But the desire to obtain general enunciations of theorems without exceptional cases has led mathematicians to employ entities of ever-ascending types of elaboration. These entities are not created by mathematicians, they are employed by them, and their definitions should point out the construction of the new entities in terms of those already on hand. The real numbers, which include irrational numbers, have now to be defined. Consider the serial arrangement of the rationals in their order of magnitude. A real number is a class (α, say) of rational numbers which satisfies the condition that it is the same as the class of those rationals each of which precedes at least one member of α. Thus, consider the class of rationals less than 2_r; any member of this class precedes some other members of the class—thus $1/2$ precedes $4/3$, $3/2$ and so on; also the class of predecessors of predecessors of 2_r is itself the class of predecessors of 2_r. Accordingly this class is a real number; it will be called the real number 2_R. Note that the class of rationals less than or equal to 2_r is not a real number. For 2_r is not a predecessor of some member of the class. In the above example 2_R is an integral real number, which is distinct from a rational integer, and from a cardinal number. Similarly, any rational real number is distinct from the corresponding rational number. But now the irrational real numbers have all made their appearance. For example, the class of rationals whose squares are less than 2_r satisfies the definition of a real number; it is the real number $\sqrt{2}$. The arithmetic of real numbers follows from appropriate definitions of the operations of addition and multiplication. Except for the immediate purposes of an explanation, such as the above, it is unnecessary for mathematicians to have separate symbols, such as 2, 2_r and 2_R, or $2/3$ and $(2/3)_R$. Real numbers with signs ($+$ or $-$) are now defined. If a is a real number, $+a$ is defined to be the relation which any real number of the form $x + a$ bears to the real number x,

and $-a$ is the relation which any real number x bears
to the real number $x + a$. The addition and multiplication of these "signed" real numbers is suitably defined,
and it is proved that the usual arithmetic of such numbers follows. Finally, we reach a complex number of
the nth order. Such a number is a "one-many" relation
which relates n signed real numbers (or n algebraic
complex numbers when they are already defined by this
procedure) to the n cardinal numbers $1, 2 \ldots n$ respectively. If such a complex number is written (as
usual) in the form $x_1e_1 + x_2e_2 + \ldots + x_ne_n$, then this
particular complex number relates x_1 to 1, x_2 to 2, $\ldots$
x_1 to n. Also the "unit" e_1 (or e_s) considered as a number of the system is merely a shortened form for the
complex number $(+1)\ e_1 + 0e_2 + \ldots + 0e_n$. This last
number exemplifies the fact that one signed real number, such as 0, may be correlated to many of the n cardinals, such as $2 \ldots n$ in the example, but that each cardinal is only correlated with one signed number. Hence
the relation has been called above "one-many." The
sum of two complex numbers $x_1e_1 + x_2e_2 + \ldots + x_ne_n$
and $y_1e_1 + y_2e_2 + \ldots y_ne_n$ is always defined to be the
complex number $(x_1 + y_1)e_1 + (x_2 + y_2)e_2 + \ldots + (x_n
+ y_n)e_n$. But an indefinite number of definitions of the
product of two complex numbers yield interesting results.
Each definition gives rise to a corresponding algebra of
higher complex numbers. We will confine ourselves here
to algebraic complex numbers—that is, to complex numbers of the second order taken in connexion with that
definition of multiplication which leads to ordinary
algebra. The product of two complex numbers of the
second order—namely, $x_1e_1 + x_2e_2$ and $y_1e_1 + y_2e_2$, is in
this case defined to mean the complex $(x_1y_1 - x_2y_2)e_1 +
(x_1y_2 + x_2y_1)e_2$. Thus $e_1 \times e_1 = e_1$, $e_2 \times e_2 = -e_1$, $e_1 \times
e_2 = e_2 \times e_1 = e_2$. With this definition it is usual to omit
the first symbol e_1, and to write i or $\sqrt{-1}$ instead of
e_2. Accordingly, the typical form for such a complex
number is $x + yi$, and then with this notation the above-
mentioned definition of multiplication is invariably
adopted. The importance of this algebra arises from the

fact that in terms of such complex numbers with this definition of multiplication the utmost generality of expression, to the exclusion of exceptional cases, can be obtained for theorems which occur in analogous forms, but complicated with exceptional cases in the algebras of real numbers and of signed real numbers. This is exactly the same reason as that which has led mathematicians to work with signed real numbers in preference to real numbers, and with real numbers in preference to rational numbers. The evolution of mathematical thought in the invention of the data of analysis has thus been completely traced in outline.

Definition of Mathematics.—It has now become apparent that the traditional field of mathematics in the province of discrete and continuous number can only be separated from the general abstract theory of classes and relations by a wavering and indeterminate line. Of course a discussion as to the mere application of a word easily degenerates into the most fruitless logomachy. It is open to any one to use any word in any sense. But on the assumption that "mathematics" is to denote a science well marked out by its subject matter and its methods from other topics of thought, and that at least it is to include all topics habitually assigned to it, there is now no option but to employ "mathematics" in the general sense[5] of the "science concerned with the logical deduction of consequences from the general premises of all reasoning."

Geometry.—The typical mathematical proposition is: "If x, y, z . . . satisfy such and such conditions, then such and such other conditions hold with respect to them." By taking fixed conditions for the hypothesis of

[5] The first unqualified explicit statement of *part* of this definition seems to be by B. Peirce, "Mathematics is the science which draws necessary conclusions," *Linear Associative Algebra*, § i (1870), republished in the *Amer. Journ. of Math.*, vol. iv (1881). But it will be noticed that the second half of the definition in the text—"from the general premises of all reasoning"—is left unexpressed. The full expression of the idea and its development into a philosophy of mathematics is due to Russell, *loc. cit.*

such a proposition a definite department of mathematics is marked out. For example, geometry is such a department. The "axioms" of geometry are the fixed conditions which occur in the hypotheses of the geometrical propositions. The special nature of the "axioms" which constitute geometry is considered in the article GEOMETRY *(Axioms)*. It is sufficient to observe here that they are concerned with special types of classes of classes and of classes of relations, and that the connexion of geometry with number and magnitude is in no way an essential part of the foundation of the science. In fact, the whole theory of measurement in geometry arises at a comparatively late stage as the result of a variety of complicated considerations.

Classes and Relations.—The foregoing account of the nature of mathematics necessitates a strict deduction of the general properties of classes and relations from the ultimate logical premisses. In the course of this process, undertaken for the first time with the rigour of mathematicians, some contradictions have become apparent. That first discovered is known as Burali-Forti's contradiction,[6] and consists in the proof that there both is and is not a greatest infinite ordinal number. But these contradictions do not depend upon any theory of number, for Russell's contradiction[7] does not involve number in any form. This contradiction arises from considering the class possessing as members all classes which are not members of themselves. Call this class w; then to say that x is a w is equivalent to saying that x is not an x. Accordingly, to say that w is a w is equivalent to saying that w is not a w. An analogous contradiction can be found for relations. It follows that a careful scrutiny of the very idea of classes and relations is required. Note that classes are here required in extension, so that the class of human beings and the class of rational featherless bipeds are identical; similarly for relations, which

[6] "Una questione sui numeri transfiniti," *Rend. del circolo mat. di Palermo*, vol. xi (1897); and Russell, *loc. cit.*, ch. xxxviii.
[7] Cf. Russell, *loc. cit.*, ch. x.

are to be determined by the entities related. Now a class in respect to its components is many. In what sense then can it be one? This problem of "the one and the many" has been discussed continuously by the philosophers.[8] All the contradictions can be avoided, and yet the use of classes and relations can be preserved as required by mathematics, and indeed by common sense, by a theory which denies to a class—or relation—existence or being in any sense in which the entities composing it—or related by it—exist. Thus, to say that a pen is an entity and the class of pens is an entity is merely a play upon the word "entity"; the second sense of "entity" (if any) is indeed derived from the first, but has a more complex signification. Consider an incomplete proposition, incomplete in the sense that some entity which ought to be involved in it is represented by an undetermined x, which may stand for any entity. Call it a propositional function; and, if φx be a propositional function, the undetermined variable x is the argument. Two propositional functions φx and ψx are "extensionally identical" if any determination of x in φx which converts φx into a true proposition also converts ψx into a true proposition, and conversely for ψ and φ. Now consider a propositional function $F\chi$ in which the variable argument χ is itself a propositional function. If $F\chi$ is true when, and only when, χ is determined to be either φ or some other propositional function extensionally equivalent to φ, then the proposition $F\varphi$ is of the form which is ordinarily recognized as being about the class determined by φx taken in extension—that is, the class of entities for which φx is a true proposition when x is determined to be any one of them. A similar theory holds for relations which arise from the consideration of propositional functions with two or more variable arguments. It is then possible to define by a parallel elaboration what is meant by class of classes, classes of relations, relations between classes, and so on. Accordingly, the number of a class of relations can be defined, or of a class of classes, and so

[8] Cf. *Pragmatism: a New Name for some Old Ways of Thinking* (1907).

on. This theory[9] is in effect a theory of the use of classes and relations, and does not decide the philosophic question as to the sense (if any) in which a class in extension is one entity. It does indeed deny that it is an entity in the sense in which one of its members is an entity. Accordingly, it is fallacy for any determination of x to consider "x is an x" or "x is not an x" as having the meaning of propositions. Note that for any determination of x, "x is an x" and "x is not an x" are neither of them fallacies but are both meaningless, according to this theory. Thus Russell's contradiction vanishes, and an examination of the other contradictions shows that they vanish also.

Applied Mathematics.—The selection of the topics of mathematical inquiry among the infinite variety open to it has been guided by the useful applications, and indeed the abstract theory has only recently been disentangled from the empirical elements connected with these applications. For example, the application of the theory of cardinal numbers to classes of physical entities involves in practice some process of counting. It is only recently that the *succession* of processes which is involved in any act of counting has been seen to be irrelevant to the idea of number. Indeed, it is only by experience that we can know that any definite process of counting will give the true cardinal number of some class of entities. It is perfectly possible to imagine a universe in which any act of counting by a being in it annihilated some members of the class counted during the time and only during the time of its continuance. A legend of the Council of Nicea[10] illustrates this point: "When the Bishops took their places on their thrones they were 318; when they rose up to be called over, it appeared that they were 319; so that they never could make the

---

[9] Due to Bertrand Russell, cf. "Mathematical Logic as based on the Theory of Types," *Amer. Journ. of Math.*, vol. xxx (1908). It is more fully explained by him, with later simplifications, in *Principia mathematica* (Cambridge).

[10] Cf. Stanley's *Eastern Church*, Lecture v.

number come right, and whenever they approached the last of the series, he immediately turned into the likeness of his next neighbour." Whatever be the historical worth of this story, it may safely be said that it cannot be disproved by deductive reasoning from the premisses of abstract logic. The most we can do is to assert that a universe in which such things are liable to happen on a large scale is unfitted for the practical application of the theory of cardinal numbers. The application of the theory of real numbers to physical quantities involves analogous considerations. In the first place, some physical process of addition is presupposed, involving some inductively inferred law of permanence during that process. Thus in the theory of masses we must know that two pounds of lead when put together will counterbalance in the scales two pounds of sugar, or a pound of lead and a pound of sugar. Furthermore, the sort of continuity of the series (in order of magnitude) of rational numbers is known to be different from that of the series of real numbers. Indeed, mathematicians now reserve "continuity" as the term for the latter kind of continuity; the mere property of having an infinite number of terms between any two terms is called "compactness." The compactness of the series of rational numbers is consistent with quasi-gaps in it—that is, with the possible absence of limits to classes in it. Thus the class of rational numbers whose squares are less than 2 has no upper limit among the rational numbers. But among the real numbers all classes have limits. Now, owing to the necessary in exactness of measurement, it is impossible to discriminate directly whether any kind of continuous physical quantity possesses the compactness of the series of rationals or the continuity of the series of real numbers. In calculations the latter hypothesis is made because of its mathematical simplicity. But, the assumption has certainly no *a priori* grounds in its favour, and it is not very easy to see how to base it upon experience. For example, if it should turn out that the mass of a body is to be estimated by counting the number of corpuscles (whatever they may be) which go to form it, then a body with an irrational measure of mass is intrinsically impossible. Similarly, the continuity of space apparently rests upon sheer assumption unsupported by any *a priori* or experimental grounds. Thus the current applications of mathematics to the analysis of phenomena can be justified by no *a priori* necessity.

In one sense there is no science of applied mathematics. When once the fixed conditions which any hypothetical group of entities are to satisfy have been precisely formulated, the deduction of the further propositions, which also will hold respecting them, can proceed in complete independence of the question as to whether or no any such group of entities can be found in the world of phenomena. Thus rational mechanics, based on the Newtonian Laws, viewed as mathematics is independent of its supposed application, and hydrodynamics remains a coherent and respected science though it is extremely improbable that any perfect fluid exists in the physical world. But this unbendingly logical point of view cannot be the last word upon the matter. For no one can doubt the essential difference between characteristic treatises upon "pure" and "applied" mathematics. The difference is a difference in method. In pure mathematics the hypotheses which a set of entities are to satisfy are given, and a group of interesting deductions are sought. In "applied mathematics" the "deductions" are given in the shape of the experimental evidence of natural science, and the hypotheses from which the "deductions" can be deduced are sought. Accordingly, every treatise on applied mathematics, properly so-called, is directed to the criticism of the "laws" from which the reasoning starts, or to a suggestion of results which experiment may hope to find. Thus if it calculates the result of some experiment, it is not the experimentalist's well-attested results which are on their trial, but the basis of the calculation. Newton's *Hypotheses non fingo* was a proud boast, but it rests upon an entire misconception of the capacities of the mind of man in dealing with external nature.

Synopsis of Existing Developments of Pure Mathematics.—A complete classification of mathematical sciences, as they at present exist, is to be found in the *International Catalogue of Scientific Literature* promoted by the Royal Society. The classification in question was drawn up by an international committee of eminent mathematicians, and thus has the highest authority. It would be unfair to criticize it from an exacting philosophical point of view. The practical object of the enterprise required that the proportionate quantity of yearly output in the various branches, and that the liability of various topics as a matter of fact to occur in connexion with each other, should modify the classification.

Section A deals with pure mathematics. Under the general heading *"Fundamental Notions"* occur the sub-headings *"Foundations of Arithmetic,"* with the topics rational, irrational and transcendental numbers, and aggregates; *"Universal Algebra,"* with the topics complex numbers, quarternions, ausdehnungslehre, vector analysis, matrices, and algebra of logic; and *"Theory of Groups,"* with the topics finite and continuous groups. Under the general heading *"Algebra and Theory of Numbers"* occur the sub-headings *"Elements of Algebra,"* with the topics rational polynomials, permutations, etcetera, partitions, probabilities; *"Linear Substitutions,"* with the topics determinants, etcetera, linear substitutions, general theory of quantics; *"Theory of Algebraic Equations,"* with the topics existence of roots, separation of and approximation to, theory of Galois, etcetera; *"Theory of Numbers,"* with the topics congruences, quadratic residues, prime numbers, particular irrational and transcendental numbers.

Under the general heading *"Analysis"* occur the sub-headings *"Foundations of Analysis,"* with the topics theory of functions of real variables, series and other infinite processes, principles and elements of the differential and of the integral calculus, definite integrals, and calculus of variations; *"Theory of Functions of Complex Variables,"* with the topics functions of one variable and of several variables; *"Algebraic Functions and their Integrals,"* with the topics algebraic functions of one and of several variables, elliptic functions and single theta functions, Abelian integrals; *"Other Special Functions,"* with the topics Euler's, Legendre's, Bessel's and automorphic functions; *"Differential Equations,"* with the topics existence theorems, methods of solution, general theory; *"Differential Forms and Differential Invariants,"* with the topics differential forms, including Pfaffians, transformation of differential forms, including tangential (or contact) transformations, differential invariants; *"Analytical Methods connected with Physical Subjects,"* with the topics harmonic analysis, Fourier's series, the differential equations of applied mathematics, Dirichlet's problem; *"Difference Equations and Functional Equations,"* with the topics recurring series, solution of equations of finite differences and functional equations. Under the general heading *"Geometry"* occur the sub-headings *"Foundations,"* with the topics principles of geometry, non-Euclidean geometries, hyperspace, methods of analytical geometry; *"Elementary Geometry,"* with the topics planimetry, stereometry, trigonometry, descriptive geometry; *"Geometry of Conics and Quadrics,"* with the implied topics; *"Algebraic Curves and Surfaces of Degree higher than the Second,"* with the implied topics; *"Transformations and General Methods for Algebraic Configurations,"* with the topics collineation, duality, transformations, correspondence, groups of points on algebraic curves and surfaces, genus of curves and surfaces, enumerative geometry, connexes, complexes, congruences, higher elements in space, algebraic configurations in hyperspace; *"Infinitesimal Geometry: applications of Differential and Integral Calculus to Geometry,"* with the topics kinematic geometry, curvature, rectification and quadrature, special transcendental curves and surfaces; *"Differential Geometry: applications of Differential Equations to Geometry,"* with the topics curves on surfaces, minimal surfaces, surfaces determined by differential properties, conformal and other representation of surfaces on others, deformation of surfaces, orthogonal and isothermic surfaces.

This survey of the existing developments of pure mathematics confirms the conclusions arrived at from the previous survey of the theoretical principles of the subject. Functions, operations, transformations, substitutions, correspondences, are but names for various types of relations. A group is a class of relations possessing a special property. Thus the modern ideas, which have so powerfully extended and unified the subject, have loosened its connexion with "number" and "quantity," while bringing ideas of form and structure into increasing prominence. Number must indeed ever remain the great topic of mathematical interest, because it is in reality the great topic of applied mathematics. All the world, including savages who cannot count beyond five, daily "apply" theorems of number. But the complexity of the idea of number is practically illustrated by the fact that it is best studied as a department of a science wider than itself.

Synopsis of Existing Developments of Applied Mathematics.—Section B of the *International Catalogue* deals with mechanics. The heading *"Measurement of Dynamical Quantities"* includes the topics units, measurements, and the constant of gravitation. The topics of the other headings do not require express mention. These headings are: *"Geometry and Kinematics of Particles and Solid Bodies"; "Principles of Rational Mechanics"; "Statics of Particles, Rigid Bodies, Etcetera"; "Kinetics of Particles, Rigid Bodies, Etcetera"; General Analytical Mechanics"; "Statics and Dynamics of Fluids"; "Hydraulics and Fluid Resistances"; "Elasticity."* Mechanics (including dynamical astronomy) is that subject among those traditionally classed as "applied" which has been most completely transfused by mathematics—that is to say, which is studied with the deductive spirit of the pure mathematician, and not with the covert inductive intention overlaid with the superficial forms of deduction, characteristic of the applied mathematician.

Every branch of physics gives rise to an application of mathematics. A prophecy may be hazarded that in the future these applications will unify themselves into a mathematical theory of a hypothetical substructure of the universe, uniform under all the diverse phenomena.

The History of Mathematics.—The history of mathematics is in the main history of its various branches. A short account of the history of each branch will be found in connexion with the article which deals with it. Viewing the subject as a whole, and apart from remote developments which have not in fact seriously influenced the great structure of the mathematics of the European races, it may be said to have had its origin with the Greeks, working on pre-existing fragmentary lines of thought derived from the Egyptians and Phœnicians. The Greeks created the sciences of geometry and of number as applied to the measurement of continuous quantities. The great abstract ideas (considered directly and not merely in tacit use) which have dominated the science were due to them—namely, ratio, irrationality, continuity, the point, the straight line, the plane. This period lasted[11] from the time of Thales, _c._ 600 B.C., to the capture of Alexandria by the Mahomedans, A.D. 641. The mediæval Arabians invented our system of numeration and developed algebra. The next period of advance stretches from the Renaissance to Newton and Leibniz at the end of the seventeenth century. During this period logarithms were invented, trigonometry and algebra developed, analytical geometry invented, dynamics put upon a sound basis, and the period closed with the magnificent invention of (or at least the perfecting of) the differential calculus by Newton and Leibniz and the discovery of gravitation. The eighteenth century witnessed a rapid development of analysis, and the period culminated with the genius of Lagrange and Laplace. This period may be conceived as continuing throughout the first quarter of the nineteenth century. It was remarkable both for the brilliance of its achievements and for the large number of French mathematicians of the first rank who flourished during it. The next period was inaugurated in analysis by K. F. Gauss, N. H. Abel and A. L. Cauchy. Between them the general theory of the complex variable, and of the various "infinite" processes of mathematical analysis, was established, while other mathematicians, such as Poncelet, Steiner, Lobatschewsky and von Staudt, were founding modern geometry, and Gauss inaugurated the differential geometry of surfaces. The applied mathematical sciences of light, electricity and electromagnetism, and of heat,

[11] Cf. _A Short History of Mathematics,_ by W. W. R. Ball.

were now largely developed. This school of mathematical thought lasted beyond the middle of the century, after which a change and further development can be traced. In the next and last period the progress of pure mathematics has been dominated by the critical spirit introduced by the German mathematicians under the guidance of Weierstrass, though foreshadowed by earlier analysts, such as Abel. Also such ideas as those of invariants, groups and of form have modified the entire science. But the progress in all directions has been too rapid to admit of any one adequate characterization. During the same period a brilliant group of mathematical physicists, notably Lord Kelvin (W. Thomson), H. V. Helmholtz, J. C. Maxwell, H. Hertz, have transformed applied mathematics by systematically basing their deductions upon the Law of the conservation of energy, and the hypothesis of an ether pervading space.

BIBLIOGRAPHY.—References to the works containing expositions of the various branches of mathematics are given in the appropriate articles. It must suffice here to refer to sources in which the subject is considered as one whole. Most philosophers refer in their works to mathematics more or less cursorily, either in the treatment of the ideas of number and magnitude, or in their consideration of the alleged *a priori* and necessary truths. A bibliography of such references would be in effect a bibliography of metaphysics, or rather of epistemology. The founder of the modern point of view, explained in this article, was Leibniz, who, however, was so far in advance of contemporary thought that his ideas remained neglected and undeveloped until recently; cf. *Opuscules et fragments inédits de Leibnitz. Extraits des manuscrits de la bibliothèque royale de Hanovre,* by Louis Couturat (Paris, 1903), especially pp. 356-399, "Generales inquisitiones de analysi notionum et veritatum" (written in 1686); also cf. *La Logique de Leibnitz,* already referred to. For the modern authors who have rediscovered and improved upon the position of Leibniz, cf. *Grundgesetze der Arithmetik, begriffsschriftlich abgeleitet von Dr. G. Frege, a.o. Professor an der Univ. Jena* (Bd. i, 1893; Bd. ii, 1903, Jena); also cf. Frege's earlier works, *Begriffs-*

schrift, eine der arithmetischen nachgebildete Formel-sprache des reinen Denkens (Halle, 1879), and *Die Grundlagen der Arithmetik* (Breslau, 1884); also cf. Bertrand Russell, *The Principles of Mathematics* (Cambridge, 1903), and his article on "Mathematical Logic" in *Amer. Quart. Journ. of Math.* (vol. xxx, 1908). Also the following works are of importance, though not all expressly expounding the Leibnizian point of view: cf. G. Cantor, "Grundlagen einer allgemeinen Mannigfaltig-keits-lehre," *Math. Annal.*, vol. xxi (1883) and subsequent articles in vols. xlvi and xlix; also R. Dedekind, *Stetigkeit und irrationales Zahlen* (ist ed., 1872), and *Was sind und was sollen die Zahlen?* (ist ed., 1887), both tracts translated into English under the title *Essays on the Theory of Numbers* (Chicago, 1901). These works of G. Cantor and Dedekind were of the greatest importance in the progress of the subject. Also cf. G. Peano (with various collaborators of the Italian school), *Formulaire de mathématiques* (Turin, various editions, 1894-1908; the earlier editions are the more interesting philosophically); Felix Klein, *Lectures on Mathematics* (New York, 1894); W. K. Clifford, *The Common Sense of the Exact Sciences* (London, 1885); H. Poincaré, *La Science et l'hypothèse* (Paris, 1st ed., 1902), English translation under the title, *Science and Hypothesis* (London, 1905); L. Couturat, *Les Principes des mathématiques* (Paris, 1905); E. Mach, *Die Mechanik in ihrer Entwickelung* (Prague, 1883), English translation under the title, *The Science of Mechanics* (London, 1893); K. Pearson, *The Grammar of Science* (London, 1st ed., 1892; 2nd ed., 1900, enlarged); A. Cayley, *Presidential Address* (Brit. Assoc., 1883); B. Russell and A. N. Whitehead, *Principia Mathematica* (Cambridge, 1911). For the history of mathematics the one modern and complete source of information is M. Cantor's *Vorlesungen über Geschichte der Mathematik* (Leipzig, 1st Bd., 1880; 2nd Bd., 1892; 3rd Bd., 1898; 4th Bd., 1908; 1st Bd., *von den ältesten Zeiten bis zum Jahre* 1200, n. Chr.; 2nd Bd., *von* 1200-1668; 3rd Bd., *von* 1668-1758; 4th Bd., *von* 1795 *bis* 1799); W. W. R. Ball, *A Short History of Mathematics* (London, 1st ed., 1888, three subsequent editions, enlarged and revised, and translations into French and Italian).   (A. N. W.)

Einstein's Theory[1]

EINSTEIN'S WORK MAY be analysed into three factors—a principle, a procedure, and an explanation. This discovery of the principle and the procedure constitute an epoch in science. I venture, however, to think that the explanation is faulty, even although it formed the clue by which Einstein guided himself along the path from his principle to his procedure. It is no novelty to the history of science that factors of thought which guided genius to its goal should be subsequently discarded. The names of Kepler and Maupertuis at once occur in illustration.

What I call Einstein's principle is the connexion between time and space which emerges from his way of envisaging the general fact of relativity. This connexion is entirely new to scientific thought, and is in some respects very paradoxical. A slight sketch of the history of ideas of relative motion will be the shortest way of introducing the new principle. Newton thought that there was one definite space within which the material world adventured, and that the sequence of its adventures could be recorded in terms of one definite time. There would be, therefore, a meaning in asking whether the sun is at rest or is fixed in this space, even although the questioner might be ignorant of the existence of

[1] The articles on this subject, which appeared on January 22 and 29 (1920), summarized the general philosophical theory of the relativity of space and time and the physical ideas involved in Einstein's researches. The purpose of the present article is in some respects critical, with the object of suggesting an alternative explanation of Einstein's great achievement.

other bodies such as the planets and the stars. Furthermore, there was for Newton an absolute unique meaning to simultaneity, so that there can be no ambiguity in asking, without further specification of conditions, which of two events preceded the other or whether they were simultaneous. In other words, Newton held a theory of absolute space and of absolute time. He explained relative motion of one body with respect to another as being the difference of the absolute motions of the two bodies. The greatest enemy to his absolute theory of space was his own set of laws of motion. For it is a well-known result from these laws that it is impossible to detect absolute uniform motion. Accordingly, since we fail to observe variations in the velocities of the sun and stars, it follows that any one of them may with equal right be assumed to be either at rest or moving in any direction with any velocity which we like to suggest. Now, a character which never appears in the play does not require a living actor for its impersonation. Science is concerned with the relations between things perceived. If absolute motion is imperceptible, absolute position is a fairy tale, and absolute space cannot survive the surrender of absolute position.

So far our course is plain: we give up absolute space, and conceive all statements about space as being merely expositions of the internal relations of the physical universe. But we have to take account of two very remarkable difficulties which mar the simplicity of this theoretical position. In the first place there seems to be a certain absoluteness about rotation. The fact of this absoluteness is inherent in Newton's laws of motion, and the deducted consequences from these premises have received ample confirmation. For example, the effect of the rotation of the earth is manifested in phenomena which appear to have no connexion with extraneous astronomical bodies. There is the bulge of the earth at its equator, the invariable directions of rotation for cyclones and anti-cyclones, the rotation of the plane of oscillation of Foucault's pendulum, and the north-seeking property of the gyro-compass. The mass of evidence is decisive, and no theory which burkes it can stand as an adequate explanation of observed facts. It is not so obvious how to combine these facts of rotation with any principle of relativity.

Secondly, the ether contributes another perplexity just where it might have helped us. We might have regained the right quasi-absoluteness of motion by measuring velocity relatively to the ether. The facts of rotation could have thus received an explanation. But all attempts to measure velocity relatively to the ether have failed to detect it in circumstances when, granting the ordinary hypotheses, its effects should have been visible. Einstein showed that the whole series of perplexing facts concerning the ether could be explained by adopting new formulæ connecting the spatial and temporal measurements made by observers in relative motion to each other. These formulæ had been elaborated by Larmor and Lorentz, but it was Einstein who made them the foundation of a novel theory of time and space. He also discovered the remarkable fact that, according to these formulæ, the velocity of light in vacuous space would be identical in magnitude for all these alternative assumptions as to rest or motion. This property of light became the clue by which his researches were guided. His theory of simultaneity is based on the transmission of light signals, and accordingly the whole structure of our concept of nature is essentially bound up with our perceptions of radiant energy.

In view of the magnificent results which Einstein has achieved it may seem rash to doubt the validity of a premiss so essential to his own line of thought. I do, however, disbelieve in this invariant property of the velocity of light, for reasons which have been partly furnished by Einstein's own later researches. The velocity of light appears in this connexion owing to the fact that it occurs in Maxwell's famous equations, which express the laws governing electro-magnetic phenomena. But it is an outcome of Einstein's work that the electro-magnetic equations require modification to express the association of the gravitational and electro-magnetic fields. This is one of his greatest discoveries. The most natural deduction to make from these modified equations is that the velocity of light is modified by the gravitational properties of the field through which it passes, and that the absolute maximum velocity which occurs in the Maxwellian form of the equations has in fact a different origin which is independent of any special relation to light or electricity. I will return to this question later.

Before passing on to Einstein's later work a tribute should be paid to the genius of Minkowski. It was he who stated in its full generality the conception of a four-dimensional world embracing space and time, in which the ultimate elements, or points, are the infinitesimal occurrences in the life of each particle. He built on Einstein's foundations, and his work forms an essential factor in the evolution of relativistic theory.

Einstein's later work is comprised in what he calls the theory of general relativity. I will summarize what appear to me as the essential components of his thought, at the same time warning my readers of the danger of misrepresentation which lies in such summaries of novel ideas. It is safer to put it as my own way of envisaging the theory. What are time and space? They are the names for ways of conducting certain measurements. The four dimensions of nature as conceived by Minkowski express the fact that four measurements with a certain peculiar type of mutual independence are required to formulate the relations of any infinitesimal occurrence to the rest of the physical universe. These ways of measurement can be indefinitely varied by change of character, so that four independent measurements of one character will specify an occurrence just as well as four other measurements of some other character. A set of four measurements of a definite character which assigns a special type to each of the four measurements will be called a measure-system. Thus there are alternative measure-systems, and each measure-system embraces, for the specification of each infinitesimal occurrence, four assigned measurements of separate types, called the co-ordinates of that occurrence. The change from one measure-system to another appears in mathematics as the change from one set of variables (p_1, p_2, p_3, p_4) to another set of variables (q_1, q_2, q_3, q_4), the variables of the p-system being functions of the q-system, and *vice versa*. In this way all the quantitative laws of the physical universe can be expressed either in terms of the p-variables or in terms of the q-variables. If a suitable measure-system has been adopted, one of the measurements, say p_4, will appear to us as a measurement of time, and the remaining measurements (p_1, p_2, p_3) will be measurements of space, which are adequate to determine a point. But different measure-systems have this property of subdivision into spatial and temporal measurements according to the different circumstances of the observers. It follows that what one observer means by space and time is not necessarily the same as what another observer may mean. It is to be observed that not every change of measure-system involves a change in the meanings of space and time. For example, let (p_1, p_2, p_3, p_4) and (q_1, q_2, q_3, q_4) be the measurements in two systems which determine the same event-particle, as I will name an infinitesimal occurrence. The two measurements of time, p_4 and q_4, may be identical or may differ only by a constant; and the spatial set of the p-system, namely (p_1, p_2, p_3), may be functions of the spatial set of the q-system, namely (q_1, q_2, q_3) with q_4 excluded and *vice versa*. In this case the two systems subdivide into the same space and the same time. I will call such two systems "consentient." A measure-system which has the property for a suitable observer of thus subdividing itself I will call "spatio-temporal." I am unaware whether Einstein would accept these distinctions and definitions. If he would not I have failed to understand his theory. At the same time I would maintain them as necessary to relate the mathematical theory with the facts of physical experience.

What can we mean by space as an enduring fact, within which the varying phenomena of the universe are set at successive times? I will call space as thus conceived "timeless space." All the measure-systems of a consentient spatio-temporal set will agree in specifying the same timeless space; but two spatio-temporal systems which are not consentient specify distinct timeless spaces. A point of a timeless space must be something which for all time is designated by a definite set of values for the three spatial co-ordinates of an associated measure-system. Let (p_1, p_2, p_3, p_4) be such a measure-system, then a point of the timeless p-space is to be designated by a definite specification of values for the co-ordinates in the set (p_1, p_2, p_3), giving the same entity for all values of p_4. Furthermore, according to Minkowski's conception, the life of the physical universe can be specified in terms of the intrinsic properties and mutual relations of event-particles and of aggregates of event-particles. Our problem then is narrowed down to this: how can we define the points of the timeless p-space in terms of event-particles and aggregates of event-particles? Evidently there is but one solution. The point (p_1, p_2, p_3) of the timeless p-space must be the set of event-particles indicated by giving p_4 every possible value in (p_1, p_2, p_3, p_4), while (p_1, p_2, p_3) are kept fixed to the assigned co-ordinates of the point. Two consequences follow from this definition of a point. In the first place, a point of timeless space is not an entity of any peculiar ultimate simplicity; it is a collection of event-particles.

Years ago, in a communication[2] to the Royal Society in 1906, I pointed out that the simplicity of points was inconsistent with the relational theory of space. At that time, so far as I am aware, the two inconsistent ideas were contentedly adopted by the whole of the scientific and philosophic worlds. To say that the event-particle (p_1, p_2, p_3, p_4) occupies, or happens at, the point (p_1, p_2, p_3) merely means that the event-particle is one of the set of event-particles which is the point. The second consequence of the definition is that if the p-system and the q-system are spatio-temporal systems which are not consentient, the p-points and the q-points are radically distinct entities, so that no p-point is the same as any

[2] "Mathematical Concepts of the Material World," *Phil. Trans.*

q-point. A complete explanation is thus achieved of the paradoxes in spatial measurement involved in the comparison of measurements of spatial distances between event-particles as effected in a p-space and a q-space. The ordinary formulæ which we find in the early chapters of text-books on dynamics only look so obvious because this radical distinction between the different spaces has been ignored.

We can now make a further step and distinguish between an instantaneous p-space and the one timeless p-space. Suppose that p_4 has a fixed value, then evidently every p-point is occupied by one and only one event-particle for which p_4 has this value. This event-particle has the p_1, p_2, p_3 belonging to its p-point and also the assigned value of p_4 as its four co-ordinate measurements which specify it. It is evident, therefore, that the set of event-particles which all occur at the assigned p-time p_4 but have among them all possible spatial co-ordinates together reproduce in their mutual spatial relations all the peculiarities of the relations between the points of the timeless p-space. Such a set of event-particles form the instantaneous p-space occurring at the p-time p_4. They are the instantaneous points of the instantaneous space. Also, all the instantaneous p-spaces, for different values of p_4, are correlated to each other in pointwise fashion by means of the timeless points which intersect each instantaneous space in one event-particle. An instantaneous space of some appropriate measure-system is the ideal limit of our outlook on the world when we contract our observation to be as nearly instantaneous as possible. We may conclude this part of our discussion by noting that there are three distinct meanings which may be in our mind when we talk of space, and it is mere erroneous confusion if we do not keep them apart. We may mean by space *either* (i) the unique four-dimensional manifold of event-particles *or* (ii) an assigned instantaneous space of some definite spatio-temporal measure-system, *or* (iii) the timeless space of some definite spatio-temporal measure-system.

We now turn to the consideration of time. So long as

we keep to one spatio-temporal measure-system no difficulty arises; the sets of event-particles, which are the sets of instantaneous points of successive instantaneous p-spaces ("p" being the name of the measure-system), occur in the ordered succession indicated by the successive values of p_4 (the p-time). The paradox arises when we compare the p-time p_4 with the q-time q_4 of the spatio-temporal q-system of measurement, which is not consentient with the p-system. For now if (p_1, p_2, p_3, p_4) and (q_1, q_2, q_3, q_4) indicate the same event-particle q_4 can be expressed in terms of (p_1, p_2, p_3, p_4) where p_4 and at least one of the spatial set (p_1, p_2, p_3) must occur as effective arguments to the function which expresses the value of q_4. Thus when we keep p_4 fixed, and vary (p_1, p_2, p_3) so as to run over all the event-particles of a definite instantaneous p-space, the value of q_4 alters from event-particle to event-particle. Thus two event-particles which are contemporaneous in p-time are not necessarily contemporaneous in q-time. In relation to a given event-particle E all other event-particles fall into three classes—(1) there is the class of event-particles which precede E according to the time-reckonings of all spatio-temporal measure-systems; (2) there is the class of event-particles which are contemporaneous with E in some spatio-temporal measure-system or other; (3) there is the class of event-particles which succeed E according to the time-reckonings of all spatio-temporal measure-systems. The first class is the past and the third class is the future. The second class will be called the class co-present with E. The whole class of event-particles co-present with E is not contemporaneous with E according to the time-reckoning of any one definite measure-system. Furthermore, no velocity can exist in nature, in whatever spatio-temporal measure-system it be reckoned, which could carry a material particle from one to the other of two mutually co-present event-particles. If E_1 and E_2 be a pair of mutually co-present event-particles, then E_1 precedes E_2 in some time-systems and E_2 precedes E_1 in other time-systems and E_1 and E_2 are contemporaneous in the remaining time-systems. The prop-

erties of co-present event-particles are undeniably paradoxical. We have, however, to remember that these
paradoxes occur in connexion with the ultimate baffling
mystery of nature—its advance from the past to the future through the medium of the present.

For any assigned observer there is yet a fourth class
of event-particles—namely, that class of event-particles
which comprises all nature lying within his immediate
present. It must be remembered that perception is not
instantaneous. Accordingly such a class is a slab of nature comprised between two instantaneous spaces belonging to the spatio-temporal measure-system which
accords with the circumstances of his observation. I have
elsewhere[3] called such a class a "duration."

The physical properties of nature arise from the fact
that events are not merely colourless things which happen and are gone. Each event has a character of its own.
This character is analysable in two components:—(1)
There are the objects situated in that event; and (2)
there is the field of activity of the event which regulates
the transference of the objects situated in it to situations
in subsequent events. It is essential to grasp the distinction between an object and an event. An object is some
entity which we can recognize, and meet again; an event
passes and is gone. There are objects of radically different types, but we may confine our attention to material
physical objects and to scientific objects such as electrons.
Space and time have their origin in the relations between events. What we observe in nature are the situations of objects in events. Physical science analyses the
fields of activity of events which determine the conditions governing the transference of objects. The whole
complex of events viewed in connexion with their characters of activity takes the place of the material ether
of the science of the last century. We may call it the
ether of events.

Now the spatial and temporal relations of event-particles to each other are expressed by the existence in space

[3] "Inquiry Concerning the Principles of Natural Knowledge"
(Cambridge University Press, 1919).

(in whatever sense that term is used) of points, straight lines, and planes. The qualitative properties and relation of these spatial elements furnish the set conditions which are a necessary prerequisite of measurement. For it must be remembered that measurement is essentially the comparison of operations which are performed under the same set of assigned conditions. If there is no possibility of assigned conditions applicable to different circumstances, there can be no measurement. We cannot, therefore, begin to measure in space until we have determined a non-metrical geometry and have utilized it to assign the conditions of congruence agreeing with our sensible experience. Practical measurement merely requires practical conformity to definite conditions. The theoretical analysis of the practice requires the theoretical geometrical basis. For this reason I doubt the possibility of measurement in space which is heterogeneous as to its properties in different parts. I do not understand how the fixed conditions for measuerment are to be obtained. In other words, I do not see how there can be definite rules of congruence applicable under all circumstances. This objection does not touch the possibility of physical spaces of any uniform type, non-Euclidean or Euclidean. But Einstein's interpretation of his procedure postulates measurement in hererogeneous physical space, and I am very sceptical as to whether any real meaning can be attached to such a concept. I think that it must be a certain feeling for the force of this objection which has led certain men of science to explain Einstein's theory by postulating uniform space of five dimensions in which the universe is set. I cannot see how such a space, which has never entered into experience, can get over the difficulty.

There is, however, another way which obtains results identical with Einstein's to an approximation which includes all that is observable by our present methods. The only difference arises in the case of the predicted shifting of lines towards the red end of the spectrum. Here my theory makes no certain prediction. A particle vibrating in the atmosphere of the sun under an assigned harmonic force would experience an increase of apparent inertia in the ratio of 1 to $^3/_5 ga/c^2$, if vibrating radially, and in the ratio of 1 to $^2/_5 ga/c^2$, if vibrating transversely to the sun's radius, where a is the sun's radius, g is the acceleration due to gravity, and c is the critical velocity which we may roughly call the velocity of light. If we assume that the internal vibration of a molecule can be crudely represented in this fashion, and if we may assume that the internal forces of the molecule are not themselves affected in a compensatory manner by the gravitational field, then we may expect a shifting of lines towards the red end of the spectrum somewhere between three-fifths and two-fifths of Einstein's predicted amount—namely, a shift and a broadening. But both these assumptions are evidently very ill-founded. The theory does not require that any space should be other than Euclidean, and starts from the general theory of time and space which is explained in my work already cited.

I start from Einstein's great discovery that the physical field in the neighbourhood of an event-particle should be defined in terms of ten elements, which we may call by the typical name $J\rho\sigma$, where ρ and σ are each written for any one of the four suffixes 1, 2, 3, 4. According to Einstein such elements merely define the properties of space and time in the neighbourhood. I interpret them as defining in Euclidean space a definite physical property of the field which I call the "impetus." I also follow Einstein in utilizing general methods of transformation from one measure-system to another, and in particular from one spatio-temporal system to another. But the essence of the divergence of the two methods lies in the fact that my law of gravitation is not expressed as the vanishing of an invariant expression, but in the more familiar way by the expression of the ten elements $J\rho\sigma$ in terms of two functions of which one is the ordinary gravitational potential and the other is what I call the "associate potential," which is obtained by substituting the direct distance for the inverse distance in the integral definition of the gravitational potential. The details of the methods and other results are more suitable for technical exposition.

Acknowledgments

Thanks are due to the following copyright holders for permission to use their material:

Atlantic Monthly. "Memories," June, 1936; "The Education of an Englishman," August, 1926; "England and the Narrow Seas," June, 1927; "An Appeal to Sanity," March, 1937; "Harvard: the Future," September, 1936.

Harvard University Press. "Process and Reality," from "Symposium in Honor of the Seventieth Birthday of Alfred North Whitehead," 1932.

Philosophical Review. "Analysis of Meaning," from "Remarks," March, 1937.

Aristotelian Society & Paul, Kegan, Trench, Trubner & Co., Ltd. "Uniformity and Contingency," Proceedings of the Aristotelian Society, Volume 23, 1922-1923.

Harvard Business Review. "The Study of the Past—Its Uses and Its Dangers," Volume XI, Number 4 (July, 1933).

Stanley Technical School & Coventry & Son, Ltd. "Education and Self-Education," from An Address delivered on Founders' Day, February 1st, 1919.

Journal of the Association of Teachers of Mathematics for the Southeastern Part of England. "Mathematics and Liberal Education," Volume I, Number 1 (1912).

Universities Bureau of the British Empire. "Science in General Education" from Proceedings of the Second Congress of the Universities of the Empire, 1921.

Radcliffe College Alumnae Association. "Historical Changes" from "Women and History," *Radcliffe Quarterly,* January, 1930.

Oxford University Press. "The First Physical Synthesis" from F. S. Marvin: "Science and Civilization."

Encyclopaedia Britannica. "Axioms of Geometry," "Mathematics," "Non-Euclidean Geometry," from 11th issue.

Mind. "Indication, Classes, Number, Validation," July, 1934.

The Times Publishing Company, Limited. "Einstein's Theory," from *The Times Educational Supplement,* February 12, 1920.

Library of Living Philosophers. Paul Arthur Schlipp, Editor. "Autobio-biographical Notes," "Immortality," "Mathematics and the Good" from Volume III, *The Library of Living Philosophers,* "The Philosophy of Alfred North Whitehead," originally published by Northwestern University Press, 1941. "John Dewey and His Influence" from Volume I, *The Library of Living Philosophers,* "The Philosophy of John Dewey," originally published by Northwestern University Press in 1939.